UNDERSTANDING FARM ANIMALS

UNDERSTANDING FARM ANIMALS

By

Dr. P.R. Yadav

Lecturer
Dept. of Zoology
D.A.V. College
Muzaffarnagar (U.P.)
(India)

DISCOVERY PUBLISHING HOUSE PVT. LTD.
NEW DELHI-110 002

Published by:
Tilak Wasan
DISCOVERY PUBLISHING HOUSE PVT. LTD.
4383/4B, Ansari Road, Darya Ganj
New Delhi-110 002 (India)
Phone : +91-11-23279245, 43596064-65
Fax : +91-11-23253475
E-mail : discoverypublishinghouse@gmail.com
sales@discoverypublishinggroup.com
web : www.discoverypublishinggroup.com

***First Edition:* 2011**

***Reprinted:* 2017**

ISBN: 978-81-8356-513-4

Understanding Farm Animals

Printed at:
Infinity Imaging Systems
Delhi

Preface

The present title "Understanding Farm Animals" has been written for those students interested in careers in diverse fields of biological sciences. It provides a structured approach to learning by covering all the important topics in a uniform, systematic format. The book has been comprehensively designed incorporating recent advances in this fast moving field. It also provides accessible information on farm animals in compact form for undergraduate students in biology and related life sciences. It is intelligible to the educated layman, though it deals with some complex ideas. It is an adequate text for all the requirements of students in this area. In addition, busy lecturers who require a quick reference compendium will find it useful, particularly for tutional planning. Simple, yet hopefully clear figures and tables are provided throughout the book.

The over-riding goal of this book, and indeed of the whole *Understanding series*, is to present the essential information concering farm animals in a compact, readily accessible form which leads itself to student learning and revision. The convergence of various approaches has generated a rich panorama of detail, the significance of which we are still attempting to unraval. The present text has been written as an introduction to this rapidly growing field.

To make the work more comprehensive and informative, the author has consulted many authoritative books, research journals, abstracts, monographs etc., so there can be no claim to originality except in the manner of treatment.

The author expresses his thanks to his friends and colleagues whose continue inspirations have initiated him to bring out this book.

The author expresses his gratitude to Mr. Wasan and staff of M/s Discovery Publishing House Pvt. Ltd. for their whole hearted co-operation in the publication of this book.

In the mean time, the author will remain sincerely responsible for any shortcomings of the book and be grateful to the readers for their suggestions and constructive criticism for the continuous betterment of the book. He takes this opportunity to appeal to the readers to send their suggestions straightaway to his Publisher.

Author

Preface

Contents

1

ANIMALS AND ENVIRONMENT

There are a very few places in the world and none in the United States where the natural climate is continuously optimum for domestic animals throughout the year. Wild animals and birds survive, as they are able to move about and select a comfortable environment. Since man began to domesticate wild animals and birds he has progressively restricted their freedom of selecting the best environment.

Man is interested in livestock and poultry production in regions that are increasingly farther removed from climates that may be ideal for optimum growth and production.

The agricultural or biological engineers in collaboration with the animal and avian scientists and livestock producers must design structures and environmental control systems that are economically optimum. This may not always mean optimum in the physiologist's sense but rather in an engineer's sense-optimum economically for profitable production.

The environment is of great importance to the agriculturalist in obtaining the fullest genetic expression from domestic animals and birds. He is concerned with:

(1) How much milk, eggs or fiber will they produce in a specific environment?

(2) Would the cost of changing the environment be compensated for by a profitable increase in production?

(3) If the environment is not now ideal, what needs to be changed and how much?

(4) Have the animals or birds had time to acclimate to the climate?

(5) What is the genetic potential of the animals or birds in an ideal or optimum environment?

(6) How may the animals' or birds' behavior and the climate be related?

To answer these questions an understanding is needed of (1) the physiological and biological bases for the animal's responses to the environment, and (2) the physical aspects of the environment and their effect on the animal's heat loss or gain to the environment.

The agricultural engineer, with some understanding of the physiological principles and a thorough knowledge of the environmental factors, can then design a shelter with an environment controlled to the extent that it is economically and genetically justified for profitable animal or bird production.

ANIMAL ENVIRONMENT FACTORS

An animal's or bird's environment is the total of all external conditions that affects its development, response, and growth. Literally, it could include the equipment and type or slope of the floor as factors of the *environclent*—and they may be important factors. It is often convenient, however, to separate the factors into physical, social, and thermal.

The physical factors are such things as space, light, sound, pressure, and equipment. The social factors are the numbers of birds or animals per cage or pen, behavior and "*peck order*." The thermal factors are air temperature, relative humidity, air movement, and radiation. This book pertains primarily to the control of these thermal factors and their effect on the animal's regulation id balance of heat production and loss.

HOMEOTHERMY

Like man, domestic animals are homeothermic—that is, they attempt to maintain a constant body temperature through a balance of heat produced and lost. The allowable range of body temperature variation is but a few degrees. Where the physiologist is concerned with the animal's production of the heat, the engineer must be concerned with and understand the loss of seat as affected by the climatic factors.

The environment surrounding an animal at any particular instant influences the amount of heat exchanged between it and that environment. Consequently it influences the physiological adjustments the animal

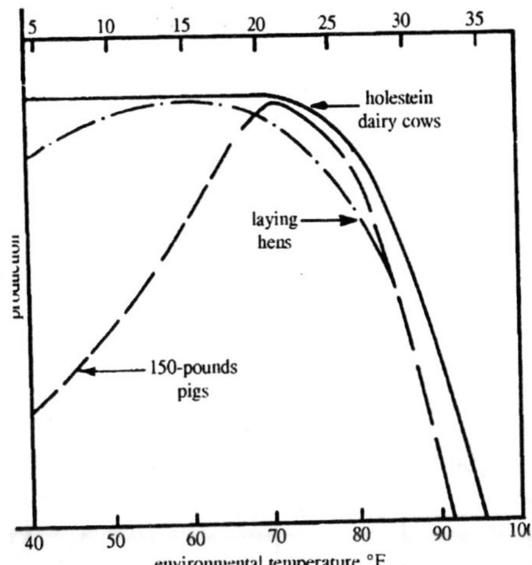

Figure 1.1 : Influence of environmental temperature on animal production trends.

must make to maintain a body heat balance. If the environment is not totally within the "*comfort zone*" of the animal the adjustments are considerable. Thus, the animal is said to be under some thermal or heat stress which would be reflected in its growth, production, and health.

DESIGN CONSIDERATIONS FOR ANIMALS

Shelter Function

The basic justification of a livestock shelter is that it should alter or modify the environment for the benefit of the animals or birds enclosed. For ninimum control the shelter would only buffer the extremes of climate to educe the peak "*stress*" on the animals or birds housed. Maximum control would include all four thermal factors.

Figure elsewhere in this chapter illustrates schematically he influence of possibly the most important factor, that of environmental temperature on animal and bird production trends. The critical effect of the temperatures in particular is emphasized by the curves of Figure elsewhere in this chapter.

Animal Production

Milk production trends, as affected by temperature, are shown more quantitatively by Figure elsewhere in this chapter. In this research, at the Psychroenergetic Laboratory at Columbia, Missouri, all environmental variables other than air temperature were controlled. At constant high temperatures, milk produce is critically affected. At low

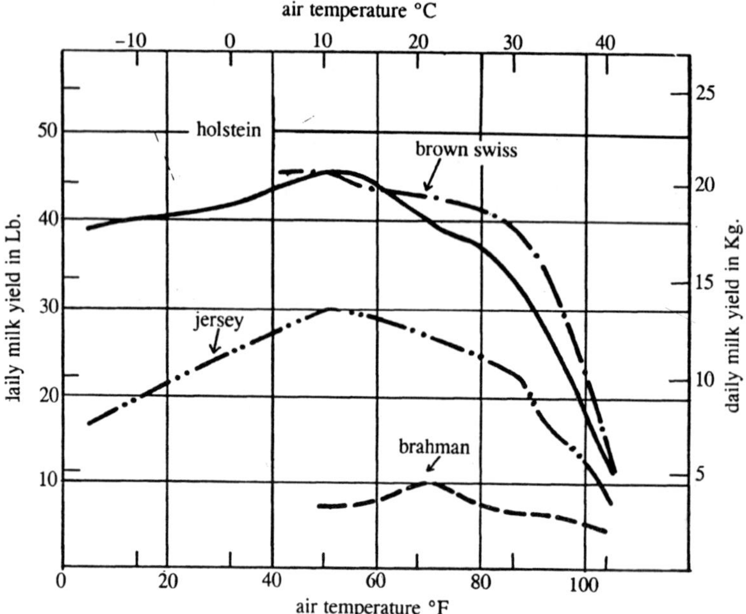

Figure 1.2: Air te mprature effects on milk yield of cows. The conditions were constant-tempreature with about 50% RH.

temperatures Jersey cows show the most decrease in milk production.

Although air temperature is the most often considered environmental factors animals, one should not conclude that the other factors are not important. Figure elsewhere in this chapter, for example, shows how high humidity as well as 1 air temperature depress milk production. Research studies of radiation air movement show, also, the critical nature of these environmental factors. These are discussed in later chapters.

Heat Production

Animal growth or product yield is the ultimate concern of the agriculturalist. Metabolic heat production' of the animal or bird is, however, often used as a key measurable index in environmental research. Figure elsewhere in this chapter shows a negative linear relationship between air temperature and heat production of dairy cattle.

Heat production in this temperature range is more consistently affected by at least this one environmental factor than milk production. Product yield and/or growth normally are not influenced by environmental factors until an accumulated stress develops under somewhat extreme conditions. Heat production is thus more directly related to the physical laws of climatic conditions.

Shelter Design

The degree of environmental control provided by a structure should be determined by the ambient environment desired for the animals and the natural climate from which they are to be protected. To date the optimum environment is not fully determined for all classes of poultry and livestock.

Even where the theoretical optimum environment is defined, the degree of environmental control may be modified for economic reasons. Economics of the cost of construction and equipment along with operation, versus production returns, dictates the extent of environmental control. As the genetic potential of animals increase and feed rations and management are more scientifically utilized, environmental control must increase.

Animals confined in a closed environment tend to alter the composition f the air by reducing the oxygen content, increasing the carbon dioxide and vapor content, and by adding ruminant gases, ammonia from feces and urine, and microscopic particles of dust from feed and bedding.

For an optimum environment designed for maximum productivity, sufficient air must be exchanged to remove detrimental levels of water vapor and ammonia, carbon dioxide, dust, and airborne microorganisms.

Under low-temperature climatic conditions, air exchange requires the addition of heat for maintenance of a desirable uniform inside environmental air temperature. Animal or bird heat may be conserved by insulated structures and utilized to warm all or part of the air exchange. Thermostatically controlled mechanical fans may be used to regulate the air exchange.

With inadequately built and insulated structures there is a tendency by managers to close them tightly during cold weather to conserve heat at the expense of sufficient air exchange. As an example of this, condemnation rates of broilers in the southern United States show a relationship between climate and housing.

Condemnations were related to time of year in Figure elsewhere in this chapter. The lowest condemnation rate in 1959 occured from May through October. The losses in the fall of 1959 continually increased from 2% to a peak of 3.8% in March of 1960. At this time parts of the southern states were exposed to low temperatures and severe snow storms.

Generally the U.S. poultry inspection data show that heavier broilers are produced in the fall, but that condemnation increases during the cold months. Broilers appear to be more efficient in feed conversion in the fall. In warm weather, smaller birds are produced but with less condemnation. These data point up implications of housing difficulty. Better environment control is needed.

Climate Modification

There are many ways of controlling or altering the thermal environment for livestock and poultry in tropical as well as cold climates. This list summarizes the various means of climate modification.

Shade
- Orientation
- Location under shade

Materials of construction
- White paint
- Water over material

Surroundings
- Grass cover
- Nearby buildings and objects
- Water directly on birds or animals
- Sprinklers or sprays
- Wallows

Partial enclosures
- Orientation
- Wind breaks
- Open buildings
- Partial walls as radiation shields
- Air movement inside
- Evaporative cooling
- Built-up manure pack
- Enclosed buildings
- Orientation
- Materials and insulation
- Location of animals or birds
- Use of water
- Mechanical ventilation
- Evaporative cooling
- Mechanical air conditioning

ENERGY TRANSFORMATION

The heat required to maintain an animal comes from the food it consumes. In the body process—metabolism—the chemical energy of the food is transformed into heat energy.

Energy is often defined as the capacity to do work. Some of the kinds or states of energy are: potential, kinetic, thermal, electrical, and radiant. Work is traditionally defined as the product of a given force acting through a given distance. Different kinds of work are mechanical, electrical, and osmotic.

As related to energy and its utilization, history is classified into three periods:

(1) before fire,

(2) use of fire and chemical energy, and

(3) control nuclear energy. The later period has greatly aided metabolic research rough the use of isotopes as tracers and the electron microscope.

Solar Energy

Agriculture involves harnessing solar energy for the production of food-stuffs. Photosynthesis by green plants and oxidation of foods by animals to yield energy for the performance of work make up this biological energy cycle. The sum total of this energy cycle far exceeds energy exchanges brought about by all man-made machines. The energy-transforming systems of living cells are also far more efficient than for any man made machine.

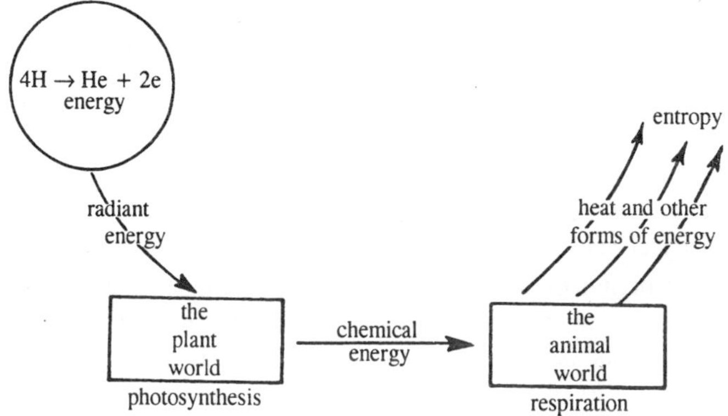

1.6. The Flow of Energy of the biological world.

Bioenergetics

Bioenergetics is defined as the study of energy transformation in living organisms. Living cells have extremely efficient and near-miraculous devices for transforming energy. The devices are molecular in dimensions and consist mainly of enzymes which are specialized proteins capable of catalyzing specific chemical reactions in the cell.

There are three stages of energy transformation in the biological world: One, photosynthesis transforms radiant energy into chemical energy with the pigment chlorophyll in green plant cells. This chemical energy is then used to form carbohydrates and other foodstuff molecules from carbon dioxide and water. Two, metabolism transforms the chemical energy of carbohydrates and other foodstuff molecules into a more useful kind of energy by oxidation in the cells of animals. Third, the energy from foodstuff oxidation is used by cells to do work either within themselves or on the environment. This may be mechanical work of muscular contraction, osmotic, electrical or the chemical work of growth

Thermodynamics

The utilization and transformation of energy by living organisms follow the basic principles of thermodynamics. The first law of energy conservation is borne out by animals in that all forms of energy are quantitatively convertible to heat. Bioorganisms are not, however, heat engines as heat is an end product rather than a motive force.

The second fundamental law of thermodynamics is that all atoms and molecules in the universe tend to seek the most random or disordered state with the least energy content. This is called entropy. The maintenance of biological complexity, therefore, requires considerable energy.

Modern technology is now demanding new modes of energy conversion besides the familiar internal combustion engines and electrical generators. Solar batteries are being designed to supply electrical power directly. The thermo-electric converter is capable of direct conversion of electrical into thermal energy and vice versa. The fuel cell is being developed to directly convert chemical energy into electrical energy.

Living organisms are providing the modern engineer with models and ideals for new methods of converting energy as well as designs for computing and communications systems. This new field is sometimes called bionics. No cheap efficient process for desalination of sea water has been perfected, yet all living cells are equipped with such molecular devices in its membranes.

Also, direct conversion of chemical energy into mechanical as occurs in the muscle is scarcely applied in modern technology. In conventional thermodynamic energy exchanges the initial and final macroscopic properties of pressure, volume, temperature, composition, and heat content must be known.

Biological energy exchange processes, although complex, can be simplified some as they usually occur at constant conditions of pressure, temperature, and volume and take place in dilute aqueous solutions.

Isothermal Process

Heat is the simplest and most familiar medium by which energy is transferred in man-made machines; however, this is not a useful way of transferring energy in biological systems.

Living organisms are essentiallyh isothermal, thus no heat can be transferred by reason or temperature difference. Cells, then, do not act as heat engines. The maximum work that be derived from a heat engine is given by the equation.

$$\text{Work} = \qquad (1.1)$$

$$= \text{heat absorbed}$$

$$T_1 \text{ and } T_2 = \text{absolute temperatures}$$

A steam engine could, for example, perform completely efficient convern of heat into work if the exhaust steam temperature was absolute zero. sere there is no temperature difference, as in a living cell, a conventional it engine has zero efficiency.

Organic compounds have a characteristic heat of combustion. It may be culculated in terms of calories as one gram molecular weight' if a substance turned completely at the expense of molecular oxygen. The equation for the combustion of glucose is

$$C_6H_{12}O_6 = 6\ CO_2 + 6\ H_2O + q$$

Thus the molar enthalpy of the combustion of glucose is

$$q = -673\ \text{kcal}^4/\text{mole}$$

Heat is produced during combustion because complex organic molecules have a large potential energy of configuration. When they are oxidized, the complex molecules are degraded to simple, stable products, such as CO_2 H_2O, which have a much lower energy content. Combustion generally brings about the greatest enthalpy change of the four common chemical fictions. Hydrolytic reactions proceed with much lower enthalpy changes.

Neutralization of H^+ and OH^- ions is also a reaction that produces heat t at a low level. In certain other ionization reactions, heat may actually be sorbed from the surroundings.

Entropy and Free Energy

If entropy is randomness or disorder, its opposite is order. The third law states that entropy of a perfect crystal or any element or compound at absolute zero temperature is zero. At absolute zero there is no thermal motion I the atoms of a perfect crystal would be in perfect order. Any state less orderly would have a finite amount of entropy.

An ionization process that absorbs heat from the surroundings does not violate the second fundamental law of themodynamics, as long as a decrease in entropy of any part of a universe is correspondingly and simultaneously accounted for by an increase in entropy of some other part. Entropy is expressed in terms of entropy units which have the dimensions of calories per mole degree.

At any given temperature, solids have relatively low entropy, liquids an intermediate amount and gases the highest. A gas at high temperature has more entropy than at low temperature due to its additional disorder and the greater thermal motion of its molecules. Thus, temperature is a significant parameter of entropy changes.

The tendency to seek the position of maximum entropy is the driving force of all processes. Heat is either given up or absorbed by the system from the surroundings to allow the system plus surroundings to reach the state of maximum entropy. The changes of heat and entropy are related by a third dimension of energy known as free energy and symbolized by G. Free energy is that component of the total energy of a system which can do work under isothermic conditions.

Entropy increases during irreversible processes and free energy decreases. Free energy is thus "useful" energy and entropy is "degraded" energy. Free energy is a useful yardstick because it can be measured more readily than entropy and is especially useful for analyzing simple biological reactions.

The fundamental equation relating changes in entropy and enthalpy to changes in free energy is

$$\Delta G = \Delta H - T\Delta S$$

where ΔG = change in free energy of the system

ΔH = heat transferred between system and surroundings

T = absolute temperature

ΔS = entropy change of the system

The free energy change of chemical reactions can, in principle, be quite accurately measured. In any chemical process, a point of equilibrium concentration, at which no further net chemical change takes place, will be reached. The constant which expresses this chemical equilibrium is

(1.2)

where A and B are the concentrations of the reactant and product respectively at equilibrium. The brackets represent the thermodynamically active masses which exist at the point of equilibrium. The equilibrium reached is a function of the drive toward minimum free energy of the reaction components.

The standard free energy change may be stated mathematically as a function of the equilibrium constant as

$$\Delta G = \Delta H - T\Delta S$$

$$\Delta G = - RTInK_{eq}$$

where R = the gas constant (1.987×10^{-3} keal/mol-deg)

T = absolute temperature

In K_{eq} = the common logarithm of the equilibrium constant

The standard free energy change is designated to take place at or near a pH 7.0.

The -standard free energy of a chemical reaction gives the maximum Mount of work that the reaction can do under isothermic conditions. This, course, can be realized only if some frictionless device exists.

When the equilibrium constant is high (meaning the reaction tends to go to complen), then the standard free energy change is negative; such a reaction prods with a decline of free energy. If K_{eq} is low, the reaction does not go far and the free energy change is positive; energy must thus be put into the system.

By analyzing and finding the final concentration of a chemical reaction the equilibrium constant may be calculated. In dilute aqueous systems, it may be assumed then that the measured concentrations are equal to the thermodynamic active masses. Having K_{eq}, the standard free energy change ty be calculated.

In both aerobic and anaerobic cells, the energy of the foodstuff molecule conserved during its oxidation, not as heat, but rather as chemical energy. The energy of the cellular oxidation reaction is,conserved in the compound enosine triphosphate which is known as ATP. The chemical energy of EP is then used to perform the chemical, mechanical, and osmotic work

the cell, as shown schematically by Figure elsewhere in this chaptrer. After the energy has been ; charged, the carrier is called ADP (adenosine disphosphate). The many emical steps in charging and discharging of the ATP system in the cell are catalyzed by enzyme systems. This is the basic principle of the cellular enzyme cycle and central to it is the substance ATP. The presence of an enzyme talyst in the system is irrelevant to the calculation as it only affects the

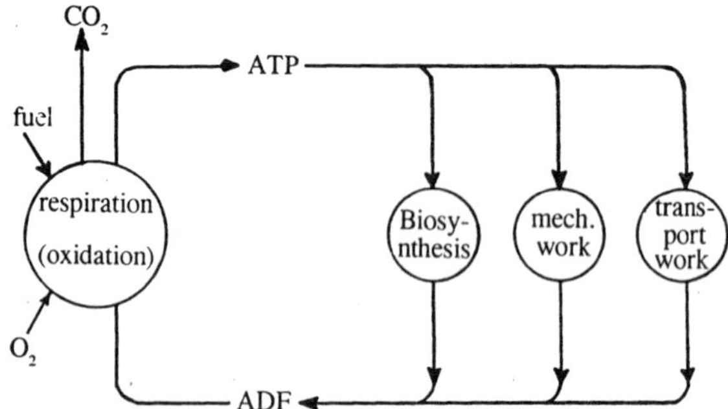

FIG. 1.7. The Transfer of Enegry in Cells By The Atp-Adp System.

rate of change and not the point attained.

The thermodynamic principles presented pertain to closed steady-state items. Living cells are, on the other hand, open systems and they exange matter with their surroundings. This complicates the mathematical alysis somewhat but it is still applicable., Living organisms also increase the entropy of the universe and function as irreversible systems. The laws of uilibrium thermodynamics do comprise a starting point for analysis of biological systems.

Biological Energy

The total amount of carbon fixed by the plant material which grows each year on the face of the earth is between 30 and 60 $\times$ 10^{15} tons. About half of this is accounted for by marine plants. For example, assume that all the fixed carbon is in the form of glucose and the formation of each mole of glucose (180 gm) requires an input of 686 kcal of solar energy.

This would require a biological energy flux of about 1 $\times$ 10^{15} kcal/yr. If this figure is corrected upward for friction losses, the total flux rises to 1 $\times$ 10^{18} kcal of captured energy per year. Thus, the biological world captures about 1/1,000 of the total.

The amount of energy estimated to be expended each year by all manmade machines is 1 $\times$ 10^{16} kcal/yr.

This is only 1/100 of the biological world energy flux. Most of the man-made machines, however, burn biologically formed fuels, such as coal, oil, and natural gas. Water power and nuclear power still provide only a small fraction of the energy requirements.

Definition of Terms

(1) *Heat* is a form of energy transmitted from one body to another by reason of temperature difference.

(2) *Temperature* refers to the ability of a body to give up or absorb heat. Arbitrary measurement scales have been developed. The *Fahrenheit* scale was based upon 0°F as the low temperature of an ice and salt mixture, 32°F as the freezing point of water, 100°F as approximate body temperature and 212 °F as the boiling point of H_2O. The *Centigrade* scale is based upon 0°C as the freezing point of water and 100°C as the boiling point of water. The relationship is

$$= (C° \times 1.8) + 32$$

(3) *Kelvin* scale is based upon the point at which the pressure of gas is zero and equals C° + 273. The *Rankine* scale is based upon the renheit degree and equals F° + 460. *fficient of thermal capacity is* the amount of heat added or taken n a substance of unit weight to change its temperature one degree

(4) *Specific heat* is the ratio of the coefficient of thermal capacity to that of .er. The specific heat for air at constant pressure = 0.24.

(5) *Latent heat* is the heat exchanged when a material changes its state of lecular aggregation without change in temperature.

Table 1.1 : Heat of Vaporiza of Water

Temperature °F	*Latent Heat, Btu/Lb*
32	1076
40	1071
60	1061
80	1048
100	1036
120	1025
200	1001
200	978
212	970

(6) Sensible heat is that which is not tied up as heat of fusion or heat of lrization. Sensible heat causes a temperature change of the

$$Q = MS(t_1 - t_2)$$

where Q = quantity of sensible heat

M = mass

S = coefficient of thermal capacity

t_1 and t_2 = temperature change of mass

Mechanical heat equivalent

1 Btu = 778 ft-Ib

1hp-hr = 2545 btu

1kw hr = 3413 Btu

1 ton of refrigeration = 12,000 Btu/hr

2

EFFECTS OF RADIATION

The surround of an animal includes the sun, sky, shade, fences, buildings, and any other object or surface that might exchange heat with the animal. Lowering the temperature of the surroundings will reduce the radiation load on an animal, or increase its radiation loss; however, this cannot always be done economically.

It is generally easier to protect animals from the hottest surroundings and increase their exposure to cool objects and surfaces. In the tropical areas of the world it is seldom economically feasible to provide closed animal structures with controlled temperature environments. *Protection* from direct and indirect solar radiation is, however, feasible and most *critically* important.

In other chapter of this book presented the basic principles of radiant heat transfer and some aspects of the long wavelength heat radiation. This chapter pertains to

(1) short *wavelength* solar radiation,
(2) some long wavelength *irradiation*, and
(3) radiation protection for domestic animals in the open environment.

By shading an animal from the sun the radiant heat load on it can be reduced from 30 to 50%. For example, on a typical August day in the *Imperial Valley* of *California*, the *radiant heat* load may be reduced from 244 to 167 Btu/(hr) (ft^2) [661.5 to 452.7 kcal/(hr) (m^2)] of animal surface by shade alone. This is equivalent to reducing the mean radiant temperature from 153 ° to 98 °F, (67.2 ° to 36.7 °C), and means that a radiant gain to the animal of about 70 Btu/(hr) (ft^2) [189.8 kcal/(hr) (m^2)] could actually be reduced to a net loss by shading.

DEFINITION OF TERMS AND INSTRUMENTATION

Mean Radiant Temperature

The mean radiant temperature (MRT) of an environment is the temperature of a uniform black enclosure with which a *globe thermometer* would exchange the same amount of energy as in the actual environment. The MRT as defined here will be strictly true only for the globe.

For a man, an animal or some other object, the MRT would be somewhat different; depending on the shape factor of the object with respect to different radiatine sources of the surroundings.

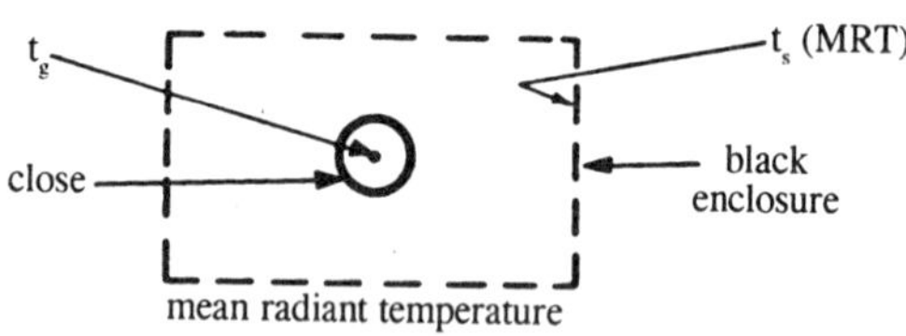

Figure 2.1: The mean radiant temperature (MRT) of an environment is the temperature of a uniform black enclosure with which a globe thermometer would exchange the same amount of energy as in the actual environment.

Radiant Heat Load

The radiant heat load (RHL) is the total radiation received by an object from all of the surrounding space. It is the spherical or wholespace irradiation of the object. Note that this is the total incoming radiation of the object and not the net exchange of radiation between the object and its surroundings.

Determination of MRT and RHL from Globe Data

Researchers studied quantitative *equilibrium* heat exchange characteristics of a 6-in. black globe and its surroundings and expressed the radiation exchange as

$$q_r = \varepsilon\,\sigma\left(T_s^4 - T_g^4\right) \tag{2.1}$$

and the convection exchange as

$$q_c = 0.169\sqrt{v}\;(t_g - t_a)$$

where t_g = temperature of the globe °F

t_a = temperature of the air °F

t_s = mean radiant temperature °F

v = air velocity in fpm

ε = emissivity of globe surface, 0.95

σ = Stefan-Boltzmann constant, 0.173×10^{-8}

$T_g = t_g + 460$

$T_s = t_s + 460$

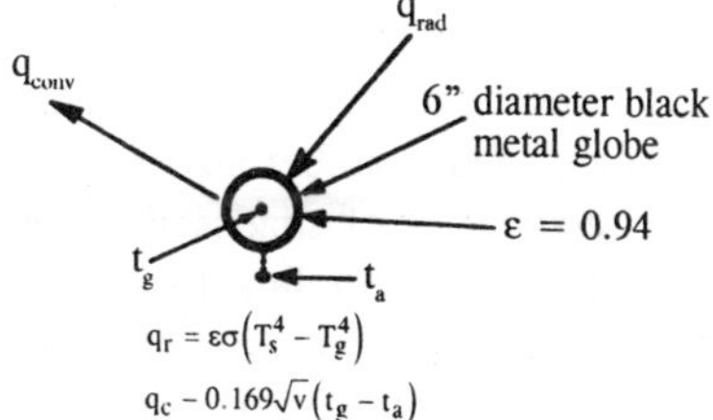

Figure 2.2 : Under steady-state conditions heat gained or lost by radiation Q_r, to or from the globe must acually equal that lost or gained by convection Q_e.

Under steady-state conditions, heat gained or lost by radiation to or from the globe must exactly equal that lost or gained by convection. Thus, equations (2.1) and (2.2) may be equated

$$t_s = 100\left[\left(T_g/100\right)^4 + 1.028\sqrt{v}\left(t_g - t_a\right)\right]^{1/4} - 460 \qquad (2.3)$$

If the equivalent uniform temperature of the surround MRT is t_s, then radiation emitted by this surround to the globe is the radiant heat load R on unit area of the globe in Btu/(hr) (ft²) as determined by the Stet Boltzmann formula

$$\text{RHL} = \sigma T_s^4 = 0.178 \times \sqrt{v}\left(t_g - t_a\right) + \sigma\, T_g^4 \qquad (2.4)$$

The mean radiant temperature t_s can be found from equation (2.3) if the air temperature, air velocity, and globe temperature are known. The RHL then can be obtained from the MRT as calculated directly by equation (2.4). Figure elsewhere in this chapter provides a *graphical* means for determining the mean radiant temperature.

Globe Thermometer

The globe thermometer is designed to provide the necessary measurements of air temperature, *globe temperature*, and *air velocity*. Test results found that 6-in. copper water trough floats (weight, 12 oz) blackened with 2 coats of flat black paint worked well as globe thermometers. *Small holes* drilled at opposite points permit insertion of 30-gage copper-constantan *thermocouples* with the injunction at the exact center of the globe. The threaded *receptacle* attached to the globe makes it convenient for hanging the globe or attaching it to a stand.

A 30-gage copper-constantant thermocouple is also attached at the receptacle so that it measures the air temperature about 2 in. from the globe. For animal environment studies the globe is placed at the approximate level of the centre of the animal considered.

Air velocity is recorded with a hot wire anemometer. Depending upon air velocity and other conditions, the globe thermometer takes a few minutes to respond.

The globe thermometer provides a reliable continuous mean radiant temperature MRT of a closed chamber environment and measurements for evaluating the *effectiveness* of shades in open environments. The globe thermometer indicates the combined effects of radiant energy, air temperature, and air velocity—three of the important factors affecting human and animal comfort. Negative radiation can be determined as readily as positive radiation.

THERMAL RADIATION SHIELDS

Shades are defined as thermal radiation shields because they may change the radiation balance of an animal, but not affect air temperature or humidity. Consideration of radiation from the sun, sky, and surroundings as related to animals is the most neglected of the *microclimate* factors at least in relation to its possible importance.

Tests in the *Imperial Valley* of *California* showed that shaded beef cattle gained 0.65 lb/day more with 143 lb less feed per 100 lb of gain than unshaded animals. In a 3-yr summary of tests in Kansas, shaded beef cattle gained significantly faster than unshaded animals. On the other hand, in tests at *Tifton*, *Georgia*, no benefits could be ascribed to the use of shades for beef cattle either in *drylot* or on *pasture*.

In tests conducted, unshaded Jersey cows in Georgia seemed to suffer more from the heat than shaded animals, but there was no evidence of any significant effect on feed and water intake, efficiency of feed conversion, or milk production. During 3 yr of tests in the Imperial Valley of California, pigs with access to a shaded wallow gained significantly faster than pigs with an unshaded wallow.

At Davis, California with a test period average of 75°F (23.9°C) no benefits were derived from shades when comparing weight gains of pigs with shaded and unshaded wallows. These findings emphasize that care must be taken in *translating* test results from one area to another.

Radiation from the upper and lower *hemispheres* is shown *graphically* by Figure elsewhere in this chapter for a particular location and time as measured with a *Gier* directional radiometer in a plane extending in an east-west direction.

It will be noted that a point in the shade, as on the right, received radiation from five sources of varying intensity as follows: (1) cool, shaded ground; (2) hot, unshaded ground; (3) horizon (a band at ground level of about 10°); (4) sky, which is quite cool; and (5) hot shade surface overhead. The unshaded point or animal on the left received radiation from only four sources. There was no cool, shaded ground, and the hot shade surface overhead was replaced by the hotter direct sun.

Method of Design

Each part of the surround radiates at an intensity depending upon its temperature and emissivity. Since any small area on the surface of the shade, ground or other part of the surround is radiating in all directions, only a part of this total energy will be *intercepted* by the animal.

Similarly, the ininitesimal areas making up the cow's surface are radiating in all directions and only a portion of the energy from each area is intercepted by the shade. The net exchange of thermal radiation between the animal and its surround depends upon the difference between the two rates of emission as well as upon the shape and relative position of the radiating surfaces.

The animal also reacts to each change in its thermal environment with changes in surface temperature and radiating rate.

For purposes of design the following assumptions are made.

(1) The heat load or *irradiation* on the animal will be calculated instead of net heat exchange.

(2) A sphere will be used as a reference surface for calculation of mean radiant temperature and radiant heat load. Although this does not truly represent an animal the relative effect is consistent. Various construction *materials*, shade height, and size may be evaluated as to the radiant seat load falling upon the small *sphere* located at a point representing the position of the animal.

(3) The surround is divided into the 5 parts as indicated by the shaded hemisphere of Figure elsewhere in this chapter.

(4) Emission rates of various parts of the *surround* must be estimated from published data, measured, or calculated.

The steps for computation of the radiation heat load under a given

Table 2.1 : Examples of energy radiated to animals under shades by various features of the surround.

Example	No. 1	No. 2	No. 3	No. 4	No. 5	No. 6	No. 7	No. 8	No. 9
Date	Sept. 1	Sept. 1	Sept. I	Sept. 1	Aug. 18	Aug. 18	Aug. 18	Aug. 22	Aug. 24
Time	11:00 A.M.	11:30 A. M.	11:30 A.M.	11:30 A. N1.	10:80 A. M.	12:30 noon	12:00 noon	12:30 P. M.	11:30 A. M.
Air temp. (°F)	106	101	101	101	96	105	105	100	104
Shade:									
Type	hay covered	galv. iron	hay over galv. iron	burlap & alum.	galv. iron	galv. iron	hay over galvn iron	hay covered	galv. iron
Size (ft)	16×24	16×24	16×24	21×21	16×24	16×24	16×24	16×24	16×42
Height (ft)	10	10	7 1/2	7	10	10	7 1/2	10	10
Radiosity (Btu/hr-ft^2):									
Ground in shadow	160	160	155	140	160	165	164	157	168
Hot ground outside shadow	210	200	200	200	260	275	275	280	275
Underside of shade	176	200	152	145	195	200	168	163	207
Horizon	205	216	216	216	215	215	215	215	240
Sky	133	145	145	145	121	130	130	142	168

(Table contd.)

(Table contd.)

Example	No. 1	No. 2	No. 3	No. 4	No. 5	No. 6	No. 7	No. 8	No. 9
Sun declination	7°57.7'	8°19'	8°19'	8°19'	13°9'	13°7.8'	13°7.8'	11°48.3'	11°8'
Sun altitude	61°30'	64°30'	63°30'	64°30'	50°35'	70°17.8'	70°17.8'	67°53'	65°27'
Sun azimuth	148°12'	158°28'	158°23'	158°23'	111°1'	180°	180°	193°51'	162°3'
Kew	0.29	0.19	0.19	0.19	0.78	0.00	0.00	0.11	0.14
Kna	0.49	0.45	0.45	0.45	0.30	0.35	0.85	0.41	0.45
Radiant heat load ($Btu/hr\text{-}ft^2$)	169	176	165	157	174	186	182	179	198

shade are as follows:

(1) Determine location of shadow with respect to the shade.

(2) Determine shape factor of each of the five principal *radiating* parts of the surround with respect to the sphere.

(3) Multiply the shape factors by the respective rates of emission.

(4) Add the five parts to obtain total radiant heat load.

Example 2.1—Determine the radiant heat load on a small sphere 3 ft above the center of the shadow. This represents the approximate center of a yearling beef animal. (If the shade is to be used for *poultry* or swine, heights more appropriate for the size of the animal should be used.) The shade is 16 by 24 ft in area and 10 ft high. The time, location and radiant heat emission (or radiosity) of parts of hemisphere will be taken from Column No. 1 of Table 2.1.

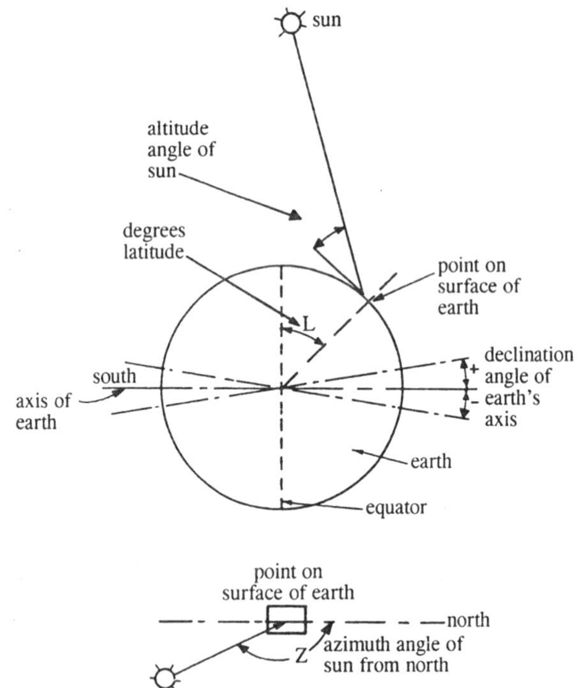

Figure 2.3 : Altitude and Azimuth angles of the sun.

Solution of Example 2.1—Step 1—The location of the shadow with relation to the shade must be determined as indicated by Figure elsewhere in this chapter. *Livestock* position themselves for *protection* from the sun by the location of the shadow and not the shade, thus this problem is to calculate the protection at the center of the shadow.

Shadow location can be measured, but for purposes of calculation a means of determining the location is necessary. The *altitude* and *azimuth* of the sun must first be determined for the specific location, season, and time of day. The position of the sun may be determined at a specific location by the equations

$$\text{Sin } \alpha = \text{Cos L Cos } \delta \text{ Cos H-Sin L Sin } \delta \quad (2.5)$$

where α = sun's altitude above horizon, deg

L = north latitude of location, deg

δ = seasonal declination of sun, deg

H = hour angle. deg (equals 15° times number of hours from solar noon, positive from 12 noon to 12 midnight

$$\text{Therefore a noon} = 90 - (L - S) \quad (2.6)$$

and

$$\text{Sin z} = \frac{\cos\delta \sin H}{\cos\alpha} \quad (2.7)$$

where z = azimuth angle of sun from due south, deg

Table 2.2 : Declination Angle d of sun, deg.

Date	*Declination Deg*
Jan. 21	-20.2
Feb. 20	-11.2
Mar. 21	0.0
Apr. 20	+11.2
May 21	+20.2
June 22	+23.45
July 23	+20.2
Aug. 24	+11 .2
Sept. 23	0.0
Oct. 23	-11.2
Nov. 23	-20.2
Dec. 22	23.4.5

For design purposes, normally the date and time of day would be selected for the maximum solar radiation (noon on June 22 in the northern *Hemisphere*) or for the maximum air temperature (usually. August).

Having selected the sun's altitude and azimuth, Figure elsewhere in this chapter may be used to determine the location of the shadow under a shade. For purposes of this problem the location of El Centro, California (34° lat. N.) will be used. On September 1, 1948 at 11:00 A.M. time the sun azimuth was 148° 12 min east and its altitude 61° 30 sec.

According to Figure elsewhere in this chapter these values for the sun azimuth and altitude indicate the shadow of a shade 1 ft high to be 0.49 ft north and 0.29 ft west of the shade corners. Multiply by 10 (the height of the shade) to obtain the shadow of the shade as 4.9 ft north and 2.9 ft west. The diagram of Figure elsewhere in this chapter indicates the relative position of the shade and shadow.

Step 2.—The shape factor of the 5 main parts of the surround must add up to unity (1.00). The shadow may be divided into sections as shown by Figure elsewhere in this chapter. The shape factors of the 4 plane areas with respect to a sphere may be determined from Figure elsewhere in this chapter.

These shape factors represent the percentage of the energy leaving the sphere which is intercepted by the surface in question. The energy leaving the plane surface and intercepted by the sphere, as for this problem, is the same. Each *plane surface* in question must have a common normal through one corner of the *rectangular* area and the center of the small sphere.

The ratios of L1/D and L_2/D are for this problem 8/3 or 2.67 and 12/3 or 4 respectively. The shape factor from Figure elsewhere in this chapter is thus 0.088. There are 4 equal parts to the shadow, thus, its total shape factor is 0.352.

The parts of the total sphere are divided into two equal hemispherical parts. The shape factor for the hot ground can then be assumed to be the difference between 0.352 and 0.500, because lower hemisphere is composed of only these 2 types of radiating areas. The upper hemisphere must be divided into three sections.

The shape factor of the 10° zone of the sky just above the horizon can be found to be 0.087 from the theorem stating that the area of a circular zone is equal to the product of the altitude and the circumference of a great circle.

The shape factor has been stated to be the fraction of the energy emitted by the reference sphere and intercepted by the second object. The shape factor of a 10° portion of a sphere of unit diameter is determined by formula :

$$\text{Shape factor} = \frac{\text{Area of Zone}}{\text{Area of Sphere}} = \frac{2\pi r \times h}{4\pi r^2}$$

$$= \frac{2\pi r^2 \sin\phi}{4\pi r^2} = \frac{\sin\phi}{2}$$

where r = radius of sphere

h = altitude of zone

ϕ = angle the zone radius makes with diameter

When ϕ, or the zone angle, is 10°, the shape factor of the sphere with respect to the zone is 0.087.

For the upper hemisphere the shape factor for the under side of the shade must be found. The 4 unequal rectangles (E, F, H, and G) of Figure elsewhere in this chapter may be evaluated with the use of the graph of Figure elsewhere in this chapter as previously done for the shadow area. The L/D ratios and partial shape factors from the curves are tabulated as follows:

Rectangle	L_1/D	L_2/D	*Shape Factor*
E	7.1/7	10.9/7	0.051
F	7.1/7	5.1/7	0.034
G	16.9/7	10.9/7	0.045
H	5.1/7	16.9/7	0.070
			0.200 Total

The total shade shape factor is 0.200. The balance of the upper hemisphere is cool sky, thus

$$0.5 - 0.087 - 0.200 = 0.213$$

as shape factor for cool sky.

Steps 3 and 4—All of the shape factors and predetermined radiosities for each area of the surround along with the summarized heat load on the sphere are given by Table elsewhere in this chapter.

Thus from Table elsewhere in this chapter the total radiant-heat load on the sphere from the surround is 168.9 Btu/(hr) (ft^2) [457.89 kcal/(hr) (m^2)].

Table 2.3: Values for shape factor, radiosity, and radiant heat load on small sphere, used for solving exmaple 2.1

Section of Surround	*Shape Factor of Surround*	*Radiosity Btu/(hr) (ft^2)*	*Heat Load on Sphere Btu/(hr) (ft^2)*
Shadow	0.354	160	56.4
Hot ground	0.148	410	31.1
Horizon	0.087	205	17.8

Shade	0.200	176	35.4
Cool sky	0.213	133	28.4
	1.000		168.9
			Btu/(hr) (ft^2)

INTERPRETATIONS OF SHADE DESIGN

A "*perfect*" shade would be one that provided 100% shadow in the lower hemisphere and 0% shade surface in the upper hemisphere. If we assume the 10° lower horizon would still be in evidence, the small sphere would have the following heat load from a perfect shade:

Section of Surround	*Shape Factor of Surround*	*Radiosity Btu/(hr) (ft^2)*	*Heat Load on Sphere Btu/(hr) (ft^2)*
Shadow	0.500	160	80.0
Horizon	0.087	205	17.8
Cool sky	0.413	133	54.9
			154.7
			Btu/(hr) (ft^2)

An animal unshaded would received radiant energy from three sources: (1) sun and sky, (2) *unshaded ground*, and (3) *horizon*. The total for the location of the example problem would be about 244 Btu/(hr) (ft^2) [661.48 kcal/(hr) (m2)].

Shade Orientation

The east-west *orientation* of the shade provides more possible exposure to the cool northern sky for more of the animals. The shadow under a north-south axis shade moves over more ground. Thus, some prefer it for better ground drying even though the thermal load is somewhat higher.

Shade Size

Increasing the size of the shade increases the shadow shape factor and reduces that of the hot ground for the lower hemisphere. The radiant heat load, however, increases for the upper hemisphere with the *hotter shade material* replacing the cool sky. Figure elsewhere in this chapter indicates little effect of shade size sere corroded galvanized iron was used at El Centro, California.

lade Height

The higher the shade the more cool sky the animal becomes exposed to. The shadow size does not change with increasing height but it does move faster and the shaded ground may not always be as cool.

With any prectable air velocity though, the hot ground cools within a few minutes after being shaded. Figure elsewhere in this chapter also shows the significance of shade height. *Cattle* have been observed to show a preference for the higher shades.

Animal Height and Location Under Shade

The lower the animal to the ground the lower the radiant thermal load. By being lower it sees a larger portion of cool shadow and less hot ground in e lower hemisphere. In the upper hemisphere it sees more cool sky and less hot shade material. Figure elsewhere in this chapter shows the effect of sphere location at 1, 2, and 4 ft above the ground.

The northern edge of the shadow is preferred with respect to radiant heating. Wind velocity and direction can, however, range the total heat balance effect on the animal.

Improving Shades

Lowering the temperature of the ground in the immediate vicinity of the will also decrease the radiant heat load. Rescue grass has been found to average 85 Btu/(hr) (ft^2) 1230.44 kcal/(hr) (ml)] lower radiosity than the hard bare ground of the corral.

Shade Cover Materials

Table elsewhere in this chapter indicates the shade radiosity is lowered by covering galvanised iron with hay. Radiosity is thus reduced from 200 to 168 Btu/(hr) (ft^2) [542.2 to 457.88 kcal/(hr) (m^2)]. If the example 9.1 hade is raised 4 ft, located in a pasture and the shade covered with hay, the adiant heat load to the animal may be reduced 22 Btu/(hr) (ft^2) [59.64 cal/(hr) (m^2)]. For 50 ft^2 of animal surface this is a saving of 1,100 Btu/hr or the equivalent latent heat of 1 lb of water evaporated from the lungs of he animal.

Some of the properties of materials which affect their efficiency as open hades are: (1) transmissivity and reflectivity of upper and lower surface for adiant energy of all wavelengths; (2) thermal diffusivity (including heat capacity, density, and thermal conductivity); (3) nature of the surface for heat transmission by convection; and (4) "solidness" or percentage of area that is open.

Numerous artificial materials were tested under typical shade design conditions at Davis, California. Comparative effectiveness of the materials is presented in Table elsewhere in this chapter. These results are based upon black-globe thermometer temperature, air dry-bulb temperature, and air velocity observations in the sun and at the center of the shadow cast by thin shades 8 by 12 L ft high with the globe 18 in. above ground. The ground w as alw ays *freshly* plowed and black, thus the radiosity of the hot ground could be reduced with an improved location.

Three or four shades were tested at the e time and one was always the new, bright, corrugated aluminum, (0.019 in. thick), for comparative purposes. The effective value was

$$E = \frac{RHL(Sun) - RHL(Sample)}{RHL(Sun) - RHL(Aluminium)}$$

Hay

This was found to be one of the best materials for a shade. It is attributed he material's relatively high insulating value, low reflection on the bottom surface, and good surface characteristics for losing heat by convection he air. Hay has *structural disadvantages*, such as poor protection from rain, possible harbor for pests, and heavier dead load than some other mails.

Aluminium

nugh aluminum is generally regarded as a good shade material it can roved by painting the top white and the bottom black.

Galvanised Steel

When galvanised steel was new it was only slightly less effective

Table 2.4 : Shade materials listed in descending order of effectiveness as compared with new corrugated aluminium

Material	*Treatment*	*Days observed*	*RHL*	*Effec-tiveness*
Hay	6 in. thick	1	150	1.203
Aluminum	top white, bottom black	5	160	1.103
Aluminum	top natural, bottom black	2	161	1.090
Galv. steel	top white, bottom black	5	163	1.066
Louvres (wood)	unpainted	1	164	1.060
Galv. steel	top white, bottom natural	3	165	1.053
Aluminum	top white, bottom natural	4	165	1.049
Neoprene nylon	top aluminum, bottom black	1	165	1.048
Louvres, wood	top unpainted, bottom black	1	166	1.042
Polyethylene, 8 mil film	black, double layer, 2 spacing	2	166	1.036

(Table contd.)

(Table Contd.)

Material	*Treatment*	*Days observed*	*RHL*	*Effec-tiveness*
Plywood, 3/8 in. thick	top white, bottom unpainted	2	167	1.031
Plywood, 3/8 in. thick	unpainted	2	167	1.030
Plywood, 1/4 in. thick	unpainted	3	167	1.030
Polyethylene, laminated	top white, bottom black	6	167	1.028
Neoprene nylon	top aluminum, bottom black	3	168	1.022
Aluminum	standard	20	170	1.000
Aluminum	1 yr old, unpainted	1	171	0.994
Galv. steel	new unpainted	5	171	0.992
Galv. steel	1 yr old, unpainted	1	172	0.985
Louvres, wood	black top, black bottom	1	173	0.970
Aluminum	10 yr old, unpainted	2	173	0.969
Asbestos board	1/8 in. thick, natural color	2	174	0.956
Building paper	aluminum coated	2	175	0.950
Hardboard	1/8 in. thick, plain (masonite)	2	176	0.942
Neoprene nylon	thin, black	1	176	0.940
Neoprene nylon	thick, black	3	177	0.933
Snow fence	double layer, no openings	1	177	0.933
Saran shade cloth	(92% solid)	1	177	0.926
Aaran pool cover cloth	—	1	181	0.889
Polyethylene, 8 mil film	black	1	183	0.868
Saran shade cloth	(92% solid)	1	187	0.839
Snow fence	double layer, crisscrossed	1	188	0.823
Polyethylene 8 mil film	translucent	1	193	0.774
Polyethylene 4 mil film	translucent	1	202	0.677
Snow fence	single layer	1	211	0.589

than aluminum as a shade material. Corrosion with age may decrease its top side reflectance but would tend to improve the lower surface.

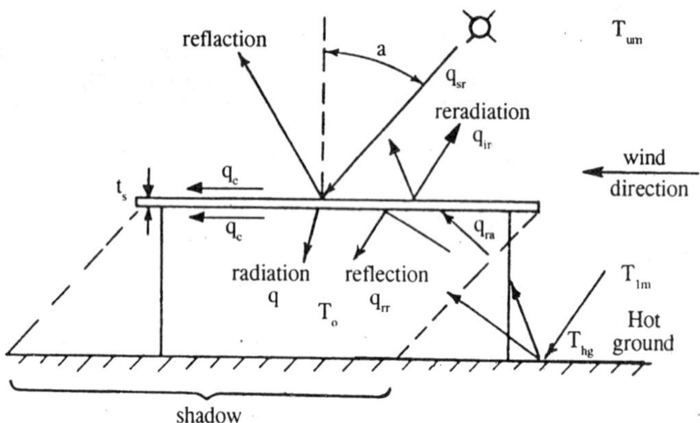

Figure 2.4 : Energy balance of a horizontal shade constructed of a thin metallic material.

UNDER SURFACE OF SHADE MATERIAL

From Table elsewhere in this chapter it is evident that black undercoatings with high emissivities are most effective as shade material. Table elsewhere in this chapter shows furthermore that the radiosity of the hot ground outside of the shadow area is as high or higher than for the underside of the different shade materials indicated.

For some materials and conditions there was over 100 Btu/(hr) (ft^2) [271.1 kcal/(hr) (ml)] difference. This was particularly true for roofing materials that were insulated with hay, and the bottom surfaces were maintained essentially at air temperature.

A perfect reflecting underc-surface material would transfer a good portion of the hot ground *radiant* intensity or radiosity directly to the animal.

If the material is a thin material with a high conduction coefficient, as most metals are, both upper and lower surface temperatures would be the same. The hay, with good convective heat loss characteristics and *insulation* quality, will have temperatures nearer to the air temperature. Upper and lower surface do not *necessarily* have the same temperature.

Figures elsewhere in this chapter show some actual surface temperature and *radiant* heat loads under shades for *corrugated* aluminum with different surface coatings. Surface temperatures in Figure elsewhere in this chapter show the white top, *plain bottom* aluminum shade to be 5° or more cooler than the other 2 materials with plain upper surfaces.

White paint and the *unpainted aluminum sheet* reflect about the same amount of solar energy (about 75%), but the emissivity of the white paint, at ordinary shade material temperatures, is much greater (about 0.89) than that of the unpainted aluminum (0.11 to 0.20).

The white painted surface consequently maintained a lower temperature because of better radiation exchange with the cool sky. Since the radiation *characteristics* of the bottom surfaces were *identical*, the radiant heat load under the cooler white painted material was considerably less.

The temperature of the third shade, with black painted underside, was about the same as that of the *unpainted shade*. The radiant heat load under it was as much as 12 Btu/(hr) (ft^2) [32.5 kcal/(hr) (ml)] less per square foot of animal surface. This difference must then be attributed to a reduction of reflected energy under the black painted shade. The undesirable effect (greater emission by the black surface) was less than the desirable effect (reduced reflection of incident energy). The black surface absorbed more radiation from the unshaded ground than did the unpainted surface, but at the same time it lost heat at a greater rate by radiation to the cool shadow. Furthermore, convection cooling tends to prevent an excessive temperature rise of the black surface.

Although the black painted shade usually had about 8°F higher surface temperature than the shade with the white top, the radiant heat load beneath it was lower by as much as 7 Btu/(hr) (ft^2) [18.98 kcal/(hr) (m^2)]. This indicates that the changes caused by the high absorptivity of the undersurface, mainly the increased surface temperature, were less important than the reduction in reflected energy.

The results shown in Figure elsewhere in this chapter emphasize the desirable effects of the white top surface and black under-surface combination. The surface temperature stays down and the radiant heat load under the shade was as much as 13 Btu/(hr) (ft^2) [35.24 kcal/(hr) (m^2)] below the unpainted aluminum.

Figure elsewhere in this chapter shows the temperature and radiant heat load under a shade made of hay to be even lower than the white top, black bottom *combination* material. Note that the hay temperature (bottom side) was as much as 13 Btu/(hr) (ft^2) [35.24 kcal/(hr) (m^2)] below the plain aluminum surface temperature.

The very uneven character of the hay surface acts as a black *surface* and *absorbs* most of the irradiation from the ground. At the same time it has good *convective cooling* characteristics to hold its temperature near air temperature.

ENERGY BALANCE

A radiation energy balance for an artifical shade is shown *schematically* in Figure elsewhere in this chapter. The shade material is thin and of a good conducting metal. Thus, both upper and lower surface temperatures are the same. The energy balance is expressed *mathematically* as

$$q + q_{rr} = q_{sr} + q_{ra} - q_{ir} - 2q_c \tag{2.9}$$

where
q = under side radiation of shade
q_{rr} = under side reflection of shade
q_{sr} = incident solar radiation absorbed on top surface
q_{ra} = radiation to under side of shade from hot ground
q_{ir} = reradiation of top surface to cool sky
q_c = convective heat loss from surfaces of shade

(1) The solar radiation q_{SI} absorbed by the upper surface is

$$q_{sr} = \varepsilon_s \mathbf{I} \tag{2.10}$$

where
ε_s = absorptivity of the roofing material for short wavelength solar radiation
I = total intensity of solar radiation on the horizontal shade surface

(2) The upper surface irradiation to the upper hemisphere is

$$q_{ir} = \sigma\,\varepsilon_1\left(T_s^4 - T_{um}^4\right) \tag{2.11}$$

where
σ = Stefan-Boltzmann coefficient
ε_l = emissivity of shade material for long wavelength radiation
T_s = absolute temperature of shade surface
T_{um} = mean absolute temperature of sky and upper hemisphere

(3) The lower surface radiation absorption from the ground is

$$q_{ra} = \sigma\,\varepsilon_1\left(T_{lm}^4 - T_s^4\right) \tag{2.12}$$

where
T_{lm} = mean absolute temperature of lower hemisphere or ground

(4) The upper and lower surface convective losses are

$$2q_c = (0.95 + 0.00276v)\,(t_s - t_a) \tag{2.13}$$

where v = air velocity in ft/min

t_a = air temperature in °F

t_s = shade material surface temperature in °F

(5) The radiation reflected downward by the underside of shade material is

$$q_{rr} = \sigma (1 - \varepsilon_1) (F) (q_{rs}) \quad (2.14)$$

where $(1 - \varepsilon_1)$ = reflection coefficient

F = shape factor to areas and positions of hot ground and under shade surface

q_{ra} = reradiation from hot ground

(6) The radiation of under roof surface is

$$q = \sigma F \varepsilon_1 (T_s^4 - T_o^4) \quad (2.15)$$

where T_o = absolute temperature of the object under shade or the ground

These energy balance equations provide the means of analytically evaluating shade materials, and the effects of hot ground surfaces, wind, and cool sky temperatures.

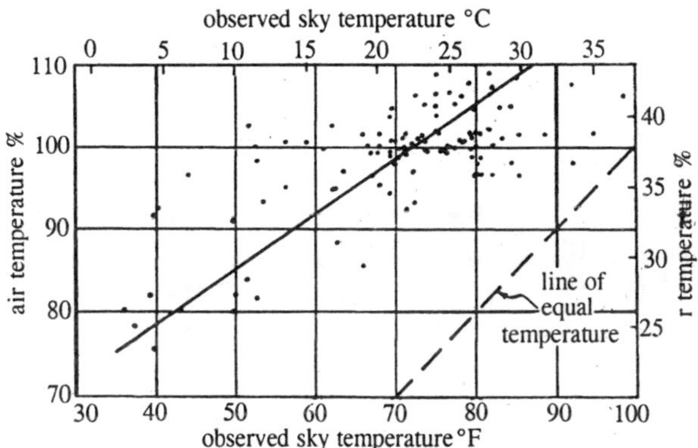

Figure 2.5 : Relation of ground air temperature to the effective north sky temperature.

The top surface of the shade material should ideally have a low *absorptivity* for short wavelength solar radiation and high emissivity for long wavelength irradiation. Table elsewhere in this chapter gives the short and long wavelength emissivity *characteristics* for some building materials and surface coatings.

Measurements of sky temperatures in the Imperial Valley of California revealed a difference of about 28°F between air and sky temperature. This is available as a cooling *sink* with air temperature at 100°F. Figure elsewhere in this chapter shows the observed sky temperature at 60° above *horizon* in the north sky with various air temperatures.

Table 2.5: Solar energy absorptyivity of various surfaces and surface emissivities at ordinary temperature.

	Shortwave Absorption	*Longwave Emission*
Metals		
Aluminum, polished	0.26	0.04—0.05
Aluminum, commercial finish, new	0.34	0.10
Chromium	0.49	0.08
Copper, polished	0.18	0.04
Copper, rolled, tarnished	0.64	0.64
Galvanised iron, new	0.85	0.13
Galvanised iron, oxidised	0.80	0.28
Zinc, pure, polished	0.46	0.02
Paints		
Porcelain enamel on steel plate (white)	0.34—0.04	0.90
Porcelain enamel on steel plate (green)	0.76	0.90
Bright aluminum new	0.40	0.43
White paint (0.017 in. on aluminum)	0.40	0.91
Black paint (0.017 in. on aluminum)	0.94—0.98	0.88
Gloss white	0.35	0.95
Light blue	0.39	0.94—0.96
Red'	0.87	0.96
Lampblack	0.98—0.97	0.96
Graphite	0.78	0.41
Building Materials		
Paper, white	0.48	0.95
Felt, roofing, bituminous	0.88	0.95
Felt, roofing, aluminized	0.38	0.50
Asbestos cement board, white	0.59	0.90
Roll roofing, green	0.88	0.91—0.97
Plaster, finished	0.35	0.93
Bricks, red	0.55	0.93

(Table contd.)

(Table Contd.)

	Shortwave Absorption	*Longwave Emission*
Glass (includes transmissivity)	0.94	0.90—0.95
Wood, planed oak		0.90
Vegetation		
Grass, high and dry	0.68	0.90
Common vegetation, field and shrubs	0.74	0.90
Grass, 80-90 % new	0.67	0.98
Tree leaves, green	0.75	0.93
Alfalfa, park green	0.97	0.95
Ground and Pavements		
Ground, dry plowed	0.78	0.90
Moist ground, 70-95%, dark	0.90	0.95
Soil, brown, dry	0.68	0.95
Soil, brown, wet	0.84	0.90
Sand, Maine, yellow and white grains	0.75	0.54
Sand, Florida, very white	0.60	0.90
Granite	0.55	0.44
Concrete	0.60	0.88
Asphalt pavement, dust-free	0.93	0.90—0.98
Snow, fresh, bright, sparkling	0.13	0.74
Earth's surface, average with cloud cover	0.57	0.90
Earth's surface, average land and sea, no clouds	0.33	0.90

It has been observed, however, that more than air temperature should be used to describe the radiating properties of the atmosphere. *Water vapor*, *gas content*, *dust particles*, and air mass all contribute to the absorption, black radiation, reflection, or transmission of radiant energy in the atmosphere. A possible method has been derived of calculating incoming radiation from observations of air temperature and vapor pressure at the ground by use of the formula

$$R = \sigma\, T^4\left(a + b\sqrt{e}\right) \tag{2.13}$$

where
R = incoming radiation (Btu/(hr) (ft^2)
σ = Stefanc-Boltzmann constant
T = temperature at the earth's surface R

$\sqrt{e}$ = mean monthly local vapor pressure of the air at the earth's surface in millibars

a = 0.55

b = 0.056

From this the effective sky temperature can be calculated by equation (2.1).

A vertical wall running east and west with the north side a perfect reflecting material, theoretically should give full exposure to the cool north sky and provide effective radiant cooling for the south as well as the north side of the animal.

An unshaded animal is *irradiated* by: (1) direct-beam solar energy from the sun; (2) diffuse sky radiation that has been scattered, reflected, and diffused out of the original beam; (3) atmospheric radiation emitted by particles or gases in the atmosphere; and (4) emitted and reflected energy from surrounding terrestrial objects.

A shaded animal, though not exposed to the direct *solar beam*, may receive solar radiation indirectly as diffuse sky radiation or as reflected energy from the ground, the shade, and surrounding objects. Terrestrial radiation comprises both "*longwave*" radiation emitted by ground objects whose temperature is close to that of air, and reflected "*shortwave*" solar or diffuse sky radiation.

"*Shortwave*" radiation of solar origin is defined as that having a wavelength of from 0.2 to 5 μ and "longwave" radiation of terrestrial origin (from the ground or *atmosphere*) between 5 and 100 μ.

Figure elsewhere in this chapter, from shortwave radiation scans in the four cardinal directions, shows the flux density normal to the line of sight under shades 6 or 12 ft high or in an unshaded area. The data are for a typical set of readings at 12 noon at El Centro, California, on September 7, 1965. Direct-beam solar energy is excluded from these data.

In this figure the diffuse shortwave energy from the zenith sky at 12 noon is 35 Btu(hr)(ft^2) while that from the 6-ft shae is about half as much, 15.6 Btu/(hr)(ft^2). When the viewing angle of the radiometers under the shade does not include the shade itself, the radiometers should show the same amount of radiation at all three locations. The slight deviation from this is due to the difference in time (about 15 min) in taking the readings at the 3 locations.

The Z-105° readings of Figure elsewhere in this chapter show the amount of radiation from the ground surrounding the shade. In the case

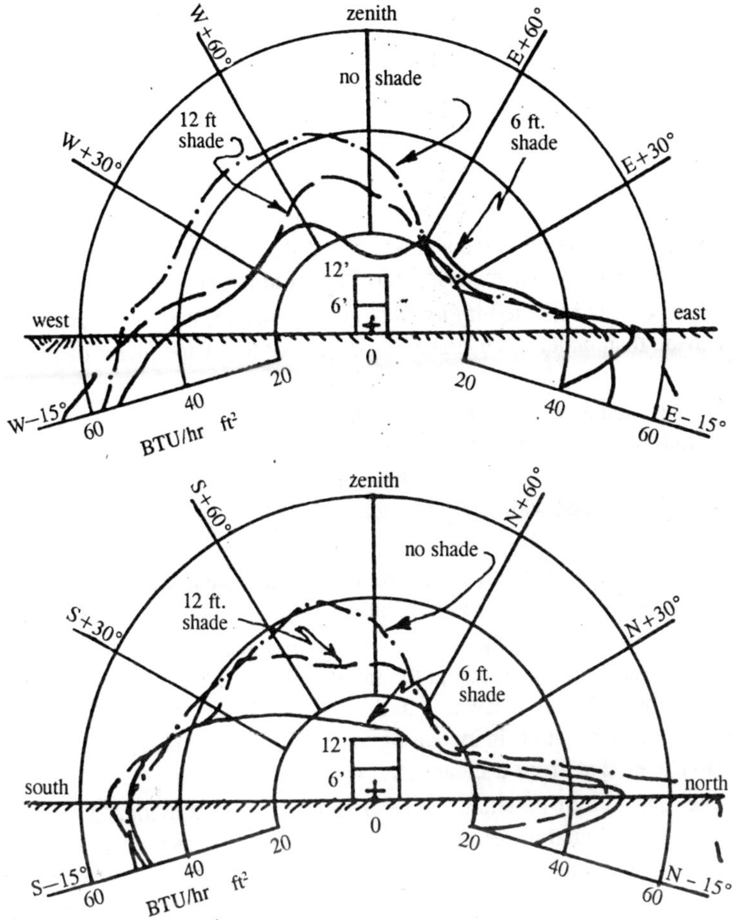

Figure 2.6: Downcoming short-wavelength radiation under shades as compared to an unshaded position.

of the diffuse *shortwave radiation*, this must be energy *reflected* from the ground. At 12 noon, both shaded and unshaded ground are included in the Z-105° measurements under the 2 shades. The radiometer sees less of the shaded ground under the 12-ft shade than under the 6-ft shade, because the shadow is farther north of the 12-ft shade; this difference is reflected in the energy measurements.

The radiometer sees no shaded ground during the unshaded readings, so the Z-105° *measurements* are higher than those under the shades in the direction of the shadow.

The graphs of Figure elsewhere in this chapter show the individual measurements for 1 set of readings. At El Centro, the percent of total radiation from the zenith that was shortwave was about 9% under the low shade, 15% under the high shade, and about 22% under the open sky; at Davis, these percentages were slightly more.

Thus, there is *considerable* shortwave energy that an animal under a shade receives from the shade material itself, since this is what the radiometer viewed in the zenith direction. This energy, originally from the sky, was reflected to the underside of the shade from the ground and then re-reflected by the material into the radiometer.

More of this reflected *diffuse* sky energy was measured under the 12-ft shade than under the 6-ft shade, because of the shape factor of the shadow.

The ground at El Centro had recently been plowed and was quite rough and cloddy; and about 23% of the radiation from the ground, in the sun, was *shortwave*. At Davis, the ground was smooth, hardcpacked, and sparsely covered with dry grass; 29% of the radiation from this was shortwave.

When the ground at either location was shaded, shortwave radiation comprised only 7% of the total radiation.

Figure elsewhere in this chapter shows longwave radiation scans made in the same directions and at the same time (12 noon) for shortwave measurements in Figure elsewhere in this chapter. It shows the influence of the shade material itself in increasing the amount of incoming longwave radiation over that from a clear sky.

The high rate of radiation from unshaded ground is evident in the comparison of ground measurements (zenith-105°) in directions toward or away from the shadow of the shades.

3

INFLUENCE OF ENVIRONMENT

Birds, like *mammals*, are *homeothermic*. The maintenance of a body temperature that varies only slightly, in spite of large variations of activity and environmental temperature, requires a *precise* control system. Heat production and heat loss must be regulated at all times. There is evidence that in birds this control is performed largely by nerve cells in the *hypothalamus*.

Numerous studies show that the *hypothalamic thermodetectors* are quite sensitive to localized thermal changes and are able to cause physiological responses. Cooling the anterior *hypothalamus* only 1°—2°C has evoked *vasoconstriction*, shivering, and increases in rectal temperature and oxygen consumption. Local heating of the anterior *hypothalamus* has induced *vasodilation*, panting, and decreases in rectal temperature, and *metabolic* rate.

Thus the hypothalamus contains *thermoregulatory* centers which process impulses from afferent paths, and initiates an integrative and coordinated neural discharge to the structures which maintain body temperature.

Some researchers assert that birds, unlike mammals, cannot be made to pant by warming the skin in the absence of an increase in the temperature of the *thermoregulatory* centers in the brain. This is based upon experiments in which the body surface was heated while cold water was circulated around the bird's neck. A close correlation was found between *hypothalamic* temperature and heart rate with respect to both the diurnal rhythms and rapid changes corresponding to level of excitement, activity, and oviposition.

The maximum daily *hypothalamic* temperature for a hen, which has laid an egg, occurred during the period of oviposition. Oviposition was also accompanied by a sharp increase in heart rate with a subsequent sharp decrease.

THERMOREGULATORY PHYSIOLOGY

The relatively high deep body temperature, the absence of sweat glands, the localised distribution of fat, and the very effective insulation provided by feathers distinguish the *thermoregulatory* physiology of birds from the other groups of homeothermic animals.

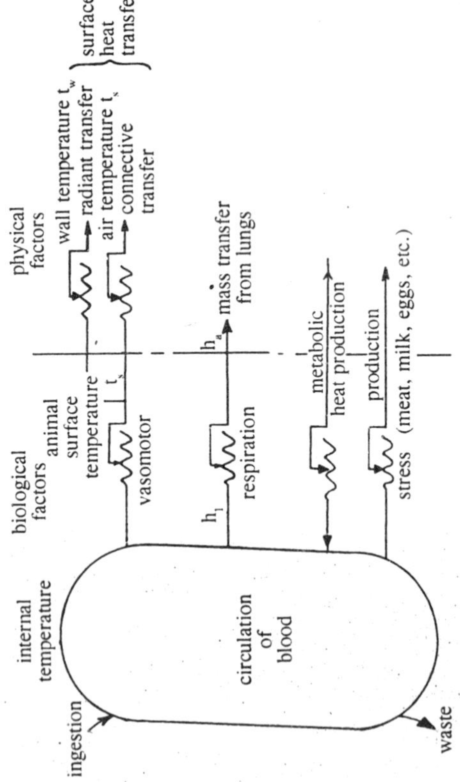

Figure 3.1: Analog Representation of homeothermic mechanisms.

Birds depend heavily on panting as a *thermoregulatory* mechanism in hot environments. An analog representation of a chicken is shown by Figure elsewhere in this chapter. The importance of respiratory and *vasomotor mechanisms* for homeothermic control is noted.

Respiratory Apparatus

The respiratory apparatus of birds is somewhat different from that of mammals. It consists of the lungs and the air passages leading to and from them. These air passages are comprised of nasal cavities, *pharynx*, *trachea*, *syrinx*, the *bronchi*, and their ramifications. The air sacs and certain of the bones of the body are also *pneumatic*.

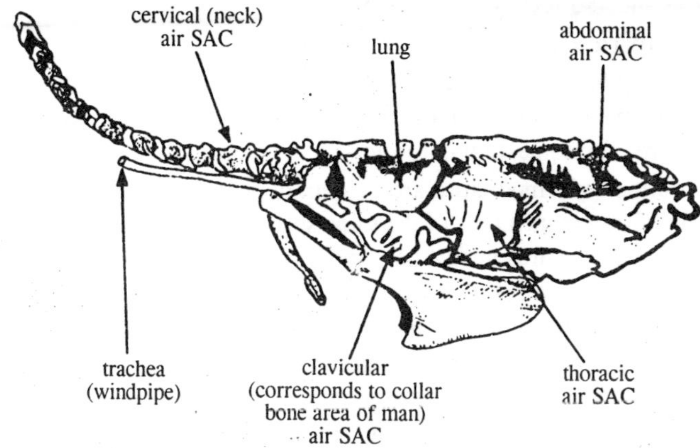

Figure 3.2: The respiratory system of a tukey which is similar to that of a chicken.

The lungs are small and are attached to the ribs of the *thorax*. The pulmonary *diaphragm* is incomplete. Some researchers consider the lungs passive in action, capable of *dilating* and contracting only when the ribs and pulmonary diaphragm do so.

Exposure to high ambient temperature, and the resulting hyperthermia, produce panting. With increasing body temperature, *respiration* rate increases and reaches a peak of 140-170 at body temperature of 43.3° to 44.4°C (110° to 112°F).

The direction of air movement in the lungs, the air sacs, and in the *primary*, *secondary*, and *tertiary* bronchi during inspiration and expiration has been studied by many but is not completely established. *Comparative partial* pressures of oxygen and carbon-dioxide in various places indicate that the inspired air first goes to the *abdominal* air sac and the dorsal bronchi.

Tidal volume, or the amount of air inspired or expired in a normal respiration, varies with respiration rate. Normal respiration rate would be about 37 per min with a tidal volume of 15.4 ml and a minute volume of 554 ml. Under hyperthermic conditions the respiration rate might increase to 138 per min, with a tidal volume of 11.3 ml and a minute volume of 1,565. Figure elsewhere in this chapter shows mean changes in respiratory rate.

Air sacs increase pulmonary ventilation and the exchange of gases in the lungs, but they are not indispensable structures to the respiratory apparatus. Many workers believed that the air sacs decreased the specific gravity of the body and thus facilitated flying but this has been discredited.

Body Temperature

The deep body temperature of birds is higher than that of the other group of homeothermic animals, the mammals. Table 10.1 gives body temperatures for some of both groups.

Variations of temperature are evident between different species of birds, and different members of the same species may also have different deep body temperatures under the same conditions. The temperature of an individual bird varies widely under different conditions.

The body temperatures of hatched chicks are lower than those for mature birds, but they increase progressively until the adult levels are reached at an age of about 20 days. Figure 10.4 shows this graphically. The increase in the deep body temperature with age appears to be associated with the growth of the plumage and with the increase in the rate of heat production.

The body temperature of birds increases after the ingestion of food, and when the plane of nutrition is increased. Conversely, starvation leads to a decrease in the body temperature. There is a diurnal variation of body temperature associated with the activity of the birds, as well as food intake periods. This body temperature rhythm appears to also be associated with heat production. Light may also play a part in the diurnal rhythm of body temperature.

Response to Changes of Environmental Temperature

For birds as with domestic animals there exists a range within which changes in environmental temperature are associated with little or no change in heat production. This is known as the thermoneutral range. Within this range, the body temperature is regulated by physical variations of heat loss. For environmental temperatures above and below

Table 3.1: Deep body temperatures of some adult birds and mammals.

Birds			*Mammals*		
	Temperature			*Temperature*	
Species	*°F*	*°C*	*Species*	*°F*	*°C*
Domestic duck	107.8	44.1	Mouse	100.2	37.9
Domestic goose	106.3	41 .3	Rat	98.3	36.8
Domestic turkey	106.2	41.4	Rabbit	104.0	38.9
Chicken	107.5	41.9	Pig	101.4	38.6
Domestic pigeon	108.0	44.4	Cat	97.6	36.4
English sparrow	110.3	43.5	Dog	100.8	38.2
Brown pelican	104.6	40.3	Sheep	104.3	39.0
Downy woodpecker	107.5	41.9	Opossum	94.5	34.7
American magpie	107.3	41.8	Echidna	82.6	48.1

this range, heat production must be varied. The thermoneutral range for any given bird may incease with age.

If the plane of nutrition is increased, or if the bird becomes acclimated to a lower environmental temperature, the lower limit may drop. The *thermoneutral* ranges are as follows: for chickens, 0-1 wk old, 34°-36°C (93.2°-96.8°F); for Rhode Island Red females at 5 wk, 32°-35°C (89.6°-95.0°F) and at 52 wk, 18°-24°C (64.4°-75.2°F).

Physiological Response to Cold: Hypothermia

The increase in heat production below the thermoneutral range is due mainly to shivering. The rate of heat production increase below this range is called the temperature *coefficient*. It is expressed in percentages of increase in the *standard* rate of heat production per degree C *reduction* in environmental temperature.

Its value depends to a large extent on the effectiveness of the bird's insulation. For the Rhode Island Red female at 52 wk the temperature coefficient is 1.5. In a cold environment, a chicken will reduce its *surface area*, and thereby its heat loss, by hunching up. It will also fluff out its feathers to increase insulation.

Tucking the head under the wing is an effective way of reducing heat loss. Reductions of 12% for chickens have been measured. Huddling is another means of heat *conservation*. Birds may also sit on their legs, which covers the *unfeathered* portions of their bodies. The heat loss in the chicken while standing is *reported* to be 40 to 50% greater than while sitting. Also increased activity and food intake increase heat production and loss. Figure elsewhere in this chapter shows the variation of skin and feather temperatures of hens with different *ambient* temperatures.

It will be noted that the skin temperatures of the uninsulated extremities of the chicken decrease as the *environmental* temperature decreases. In some birds there are special vascular structures (rete) in the legs which *permit* cooling of the *arterial* blood going to the feet by the cold returning venous blood.

These structures are present particularly in the legs of wading birds. It has also been found that for environmental temperatures below freezing, the *extremities* are prevented from freezing by *periodic* increases in the flow of warm blood which results in periodic increases in temperature.

This phenomenon is called "*cold vasodilation.*" The importance of the plumage of birds during exposure to cold will be noticed in Figure elsewhere in this chapter.

Physiological Responses to Heat—Hyperthermia

As the environmental temperature increases, Figure elsewhere in this chapter indicates that the skin temperatures beneath the feathered areas increase more slowly than the *unfeathered* areas. The temperature of the extremities is increased by increasing blood flow.

For cooling purposes the legs are well adapted as they have poor insulation with a large surface area-volume ratio. Figure 10.6 shows this quick rise of from 5° to 8°F in skin temperatures of extremities within the first hour.

Panting in chickens is initiated with deep body temperature between 41.0° to 43.5°C (105.8° to 110.3°F). *Respiration* rate increases as well as minute volume as indicated by Figure elsewhere in this chapter.

The tidal volume, however, decreases with increasing environmental temperature supposedly to restrict the hyperventilation to the surfaces of the *respiratory tract*. In this way, the possibility of the removal of excessive amounts of carbon-dioxide from the blood is *lessened*.

In the chicken the arterial blood pressure and the calculated total peripheral vascular resistance to blood flow are lowered during *hyperthermia*, *presumably* as a result of vasodilation in the extremities.

Increased cardiac output in blood volume then enhances the rate of blood flow through the *extremities*, the evaporating areas of the respiratory tract, and the respiratory muscles involved in panting. The plasma volume increases slightly during *hyperthermia* in spite of the evaporative loss of water. The consumption of water is, however, higher as well as the feces volume.

TOLERANCE TO HEAT AND COLD

The rectal temperatures of various White Leghorn chickens at various air temperatures are given in Figure elsewhere in this chapter. The highest environmental temperatures that birds can *tolerate* without progressive increase in their body temperature depend, among other things, upon the humidity of the air. The degree of *acclimatization* also influences heat tolerance. A lower plane of nutrition aids heat tolerance although it may cut production.

Evaporative Cooling

Dew-point temperatures appear to be critical for chickens. Test findings concluded that increasing dew-point temperature depresses feed consumption independent of dry-bulb temperature. This has *farreaching implications* and perhaps this tendency toward lower feed consumption

Table 3.2: Total feed consumption in GM as related to dry-bulb and dew-point temperature.

Dry-Bulb Temperature	*Dew Points* 60°F	66°F	74°F
90.5	44.1	8.8	3.9
81.5	55.5	33.9	49.7
74.5	60.8	47.1	14.7

with increasing dew-point temperature may explain failure of evaporative cooling to show significant production advantages in some field tests. It was further recommended that a low dew-point temperature (< RH 60%) is to be desired under all summer conditions. *Evaporative* cooling should improve the heat loss from birds but may tend to depress the *appetite* sufficiently to restrict any great improvement in *egg production*.

Four years of studies on the effect of evaporative cooling on laying chickens at Davis, California, have led to the following conclusions. (1) Improved bird "*comfort*" does not necessarily result in increased egg production, *egg weight*, *fertility*, or *hatchability*. (2) The temperature-humidity index (THI) concept of summer comfort appears useful in explaining the negative results in *California* and the positive results in Texas from evaporative cooling of caged layers. (3) Low nighttime temperatures are considered to be a *contributory* reason why evaporative cooling gave no benefits at Davis. (4) Fringe benefits of evaporative cooling are substantial enough that cooling may be used in areas where it cannot be justified from egg production response.

The temperature-humidity index THI = $0.4\ (t_{db} + t_{wb}) + 15$ appears to offer some reasonable measure of comparison between results for evaporative cooling at different locations. Research at Athens, Georgia, as at Davis, California, concluded that evaporative cooling and fogging both reduced rectal temperature of laying hens, but had little if any effect on the egg production of light breeds of chickens.

It was reported that under Texas conditions a statistically significant increase in egg production and hatchability was obtained from evaporatively cooled hens. It was also found that on light-breed hens, the pad-and-fan evaporative cooling system gave an egg production advantage of 4.2% over a package-type evaporative cooler, and 4.5% over fogging.

Table elsewhere in this chapter indicates that the climates at Athens and Davis are somewhat comparable; the maximum dry-bulb temperature at Davis is higher than at Athens, but the minimum and mean temperatures are lower. Likewise, on the THI index basis, Athens and Davis have the same average maximum value, although Davis has a considerably lower mean.

These differences are due to the large diurnal temperature cycle of Davis-a 40°F difference between day and night is not uncommon. In addition, Davis has a lower relative humidity (and wet-bulb temperature) than does Athens.

By all comparisons, except average maximum dry-bulb temperature and number of days above 100°F, Dallas is considerably warmer than Davis and Athens. At Davis, the evaporative coolers were effective in reducing temperatures so that the cooled birds were never *stressed* to any extent.

The *uncooled* birds were *undoubtedly* stressed a number of times as there were 85 days in 4 years with temperatures of 100°F or above. Severe panting was frequently observed on the uncooled birds on such days. The question arises as to why this did not affect egg production. It can only be stated that a correlation between panting and egg production has not been proven.

In addition to this researchers feel that the large diurnal temperature cycle of Davis was a modifying effect. If a chicken does not die on a hot day, the low temperature during the following night allows a recovery with possibly no consequent after effects.

Thus it appears that evaporative cooling of poultry may be expected to be beneficial only in areas where the dry-bulb temperature of THI is very high (THI 75 or above) and in addition, there is little opportunity for diurnal recovery from such heat stress.

ENERGY METABOLISM AND HEAT LOSS VS. ENVIRONMENT

The energy retained in the body is the metabolized energy and amounts to between 70 and 90% of the gross energy depending on the diet, the environmental temperature, the species of bird, and other factors. The absorption of energy from the *gastrointestinal* tract is followed, soon after its *absorption*, by an increase in heat production.

This heat is referred to variously as the heat increment, the calorigenic effect or the specific dynamic action (SDA) of the diet. The heat is thought to be derived from the *exothermic* reactions associated with the *metabolism* of the absorbed food *molecules*.

If no work is performed and if the body weight, composition, and temperature do not change, then all of the *metabolizable* energy appears as heat. It is only under these conditions that the measurement of heat production by a bird is a valid measure of the rate of its metabolism.

There is a *diurnal rhythm* of metabolism in fasting birds which is independent of the effects of food intake. The metabolic rate of the chicken is highest in the forenoon and lowest at about 8:00 P.M. When chickens are allowed to feed *ad libitum*, there is however, a diurnal rhythm of food consumption which precedes, in time, the rhythm of heat production.

Greater diurnal variations of *metabolism* are evident in young birds than adults. The reduction in the heat production at night may amount to 18 to 30% in the chicken.

When birds are exposed to environmental temperatures below their critical temperature, their heat production increases to a maximal value that may be 3 or 4 times greater than normal, as illustrated by Figure elsewhere in this chapter. The increase in heat *production* is brought about largely by *shivering*.

The metabolic output for different aged chickens as affected by environmental temperature is dramatically shown by Figure elsewhere in this chapter. The minimum tolerable temperatures for young birds is shown at the peak of the curves.

Also the lowering of the thermoneutral zone with age is evident. If the straight portion of each of the *curves* was to be extended downward to the right until it crossed the x-axis, it is significant to note that the intercept moves to the left for younger birds. This confirms the fact that as the environmental temperature is lowered below body temperature, the younger birds having less insulation must compensate more quickly with equivalent heat production to offset heat loss.

If birds could only dissipate heat in the sensible forms, (*radiation*, *convection*, and *conduction*) the heat loss curves would theoretically all meet the body temperature and zero *metabolism* points *simultaneously*.

The heat production of chickens can be calculated from oxygen consumption data with the following formula

$$Q = 3.871\ O_2 + 1.194\ CO_2 \qquad (10.1)$$

where Q = heat production in kcal

O_2 = oxygen consumption in liters

CO_2 = carbon-dioxide production in liters

The respiratory quotient (RQ) is the ratio of the volume of *carbon dioxide* produced to the volume of oxygen consumed in a given time. The RQ varies with the diet. Table elsewhere in this chapter summarizes this.

Table 3.3 : Factors for use in indirect calcrimetry in birds.

Substance	*Metabolizable Energy (Kcal/Gm)*	*Thermal Quotient (Kcal/Liter O_2)*	*Respiratory Quotient*
Carbohydrate	4.2	5.047	1.00
Protein	4.2	4.75	0.73
Fat	9.5	4.686	0.71

If the comparative amounts of *carbohydrate*, *protein*, and fat are known, an RQ can be estimated. Most accurate data are, however, obtained by measuring both O_2 and CO_2 under calorimetry coriditons.

Heat and Vapor Loss by Chickens

The heat loss of a bird as with an animal may be expressed by a version of Newton's Law of Cooling, $Q = AkU\Delta t$, as follows

$$Q = k = \frac{(t_e - t_a)}{R_t + R_F + R_A} + Q_E \qquad (3.2)$$

where Q = total heat loss by one chicken

k = an overall coefficient of heat transfer

t_c = core or deep body temperature

t_a = ambient air temperature

R_t = insulation of tissues
R_F = insulation of feathers
R_A = insulation of air
Q_E = heat lost by evaporation.

This being a general equation for 1 bird the surface area would be unity for that 1 bird. The total heat produced is dissipated into the environment by means of *conduction*, *convection*, *radiation*, and *evaporation*.

The first three modes of heat transfer are the means by which sensible heat is transferred to the environment by virtue of a temperature difference as indicated by equation. Figure elsewhere in this chapter shows some of the temperature differences that exist on the bird's various surfaces at different environmental temperatures.

With *partitional calorimetric* techniques scientists measured the sensible heat loss partition of seven birds and evaluated the resistance to the transfer of heat by convection and radiation. These calculated values are tabulated in Figure elsewhere in this chapter.

These measurements were taken with all surrounding surface temperatures at air temperature. Thus any interpretation for housing design and environmental control would only hold at *specific* times during *spring* and fall when such temperature combinations might occur.

Sensible heat loss partitioning from Leghorn layers was further studied. Based upon *calorimeter* tests of 27 birds the following equations were developed

$$Y_1 = 19.0 - 0.374X_1 + 4.81X_2 + 57.9X_3 + 0.298X_4 \quad (3.3)$$

where Y_1 = total heat production, Btu/hr
X_1 = dry-bulb temperature, °F
X_2 = bird weight, lb
X_3 = feed consumption, lb/day
X_4 = dew-point temperature, °F

$$Y_2 = -0.780 + 0.0158X_1 + 0.053X_2 - 0.00792X_4$$
$$Y_3 = +1.780 - 0.0158X_1 - 0.053X_2 + 0.00792X_4$$
$$Y_4 = +0.793 - 0.00653X_1 - 0.02685X_2 + 0.00239X_4$$
$$Y_5 = +0.989 - 0.00929X_1 - 0.0261X_2 + 0.00554X_4$$

where Y_2 = fraction of Y_1 dissipated as latent heat
Y_3 = fraction of Y_1 dissipated as sensible heat
Y_4 = fraction of Y_1 dissipated as radiant heat

Y_5 = fraction of Y_1 dissipated as convective heat
X's are same as above.

Some selected solutions of the 5 equations are presented in Table elsewhere in this chapter. Cases I and II indicate what might happen if the ambient air was saturated adiabatically. Case I assumes a constant feed *consumption* and

Table 3.4 : Solutions to Equations 1 through 5.

Temperature °F								
X_1 Dry-Bulb	X_4 Dew Point	*Bird Weight* X_2 Lb	*Feed Consump.* X_3 Lb/Day	Y_1 Btu/hr	Y_2 %	Y_3 %	Y_4 %	Y_5 %
Case I:								
90	60	3,975	0.218	28.4	38	62	24	38
80	65	3.975	0.218	33.7	18	82	32	50
70	69.5	3.975	0 218	38.8	0+	100–	39	62
Case II:								
90	60	3.975	0.218	48.4	38	62	44	35
80	65	3.975	0.174	32.5	18	82	32	50
70	69.5	3.975	0.087	35.1	0+	100–	39	62
Case III:								
90	60	3.975	0.418	28.4	38	62	24	38
90	66	3.975	0.418	30.2	33	67	2.5	44
90	72	3.975	0.e18	34.0	28	72	27	45
Case IV:								
75	60	3.975	0.327	37.1	14	86	34	54
75	66	3.975	0.418	35.8	9	91	35	56
75	72	3.975	0.109	34.5	4	96	37	59
Case V:								
90	60	3.975	0.218	28.4	38	62	24	39
80	60	3.975	0.272	33.7	22	78	31	47
70	60	3.975	0.305	38.3	6	94	37	57

Case II a somewhat more likely declining feed consumption. Case III shows a variation of increasing dew-point temperature with a high constant drybulb temperature and constant feed consumption. This is a highly unlikely situation and a misleading mathematical solution. Case

IV is a very possible situation at a warm dry-bulb temperature, increasing dew-point temperature and decreasing feed consumption. Total heat production and sensible heat production both decrease.

Case V more closely resembles the situation of fluctuating dry-bulb temperature in summer coupled with a reasonably steady dew-point temperature. It reflects a somewhat arbitrary adjustment of feed consumption. It is interesting to note the variation in the modes of transmission of sensible heat for Case V.

If the Newton Law of Cooling equation ($q_c = k A \Delta t$) is applied to birds for convective losses under the conditions of Case V, in Table 10.5 it is found at 70°F (21.1°C) dry-bulb temperature and a body temperature of 106°F (41.1 °C) that the temperature difference At, is 36°F (20°C) and k × A = 0.59 Btu/(hr) (°F). If it is assumed for comparison purposes that k x A does not change from 70° to 90°F dry-bulb temperature, the heat transfer by convection should be 21.8 Btu/hr at 70°F, 15.9 Btu/hr at 80°F, and 10.0 Btu/hr at 90°F.

Table 3.5 : Radiation cooling of laying hens.

t - °F	q_r - Btu/Hr	t — °F	Body Temp	$\frac{106 - t_1}{106 - t_2}$
70	38.3 × 0.37 = 14.1	79	106	$\frac{27}{3} = 0.73$
80	10.5	87	106	$\frac{19}{26} = 0.73$
90	6.8	94	106	$\frac{12}{16} = 0.75$

The calculated heat loss by convection in Case V found by multiplying Y, time Y,/100 is 21.8 Btu/hr at 70°F, 15.8 Btu/hr at 80°F, and 10.8 Btu/hr at 90°F. So in actuality, based on this, k × A does change but very little between 70° to 90°F as result of "feather fluffing," "wing extension," etc. Based on an assumption of 1.5 sq ft as the surface area of the average sized bird, the heat loss coefficient of the feathers would be in the order of 0.39 Btu/ (hr) (°F) (ft^2).

If the radiation cooling of Case V is examined in a similar way using the equation $q_r = F \sigma A (T_1^4 - T_2^4)$ it may be determined if the bird makes any sizable changes in its radiant effectiveness. For these calculations F is assumed as 0.97, a as 0.1714×10^{-1} Btu/ (hr) (ft^2) (°R^4), A as 1.5 ft^2 of black body surface, T, as effective radiating temperature of the bird surface in °R, and T2 as temperature of enclosed surface in °R. Calculations in Table elsewhere in this chapter show the apparent t, at which *radiation* must have orginated at the feathers to have transferred the amount of heat indicated as a product of $Y_1 \times Y_4/100$ in Case V.

These calculations indicate that the effective radiation temperature remains at the same point in the temperature gradient from body temperature to ambient temperature.

HEAT AND MOISTURE PRODUCTION OF LAYERS AND BROILERS

Total heat production of layers depends upon breed, production level, nutrition level, size and age of hen, and environmental temperature. The total heat production per pound of live weight was reported for

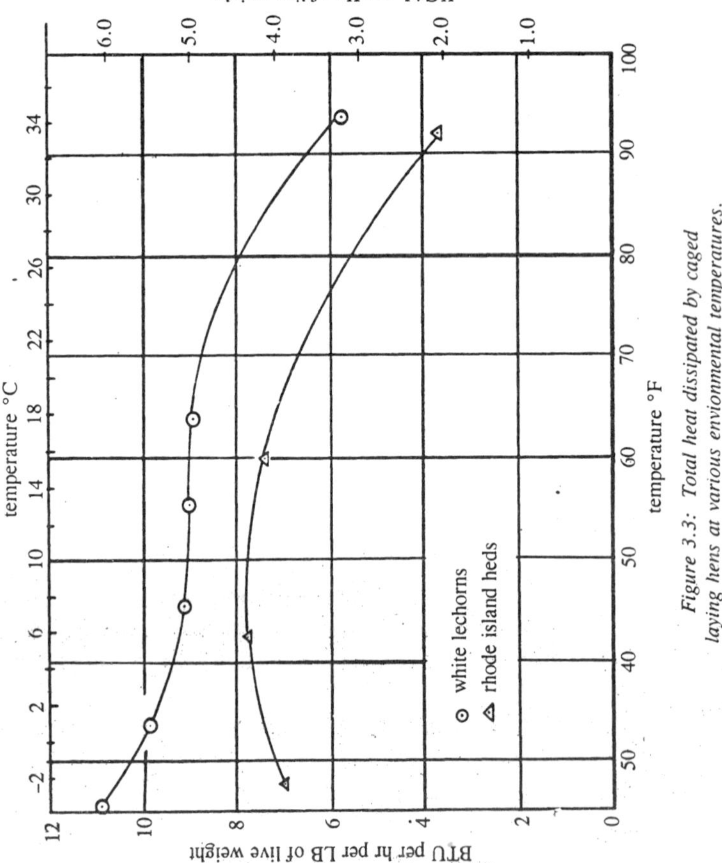

Figure 3.3: Total heat dissipated by caged laying hens at various envionmental temperatures.

White Leghorns and Rhode Island Red hens at temperatures of 25° to 94°F (-3.9° to 34.4°C) and is presented in Figure elsewhre in this chapter.

It is noted that heat production for the light breed, White Leghorn hens, is fairly consistent at 9 Btu/lb between the environmental temperatures of 45° and 65°F (7.2° and 18.3°C). In using these data for environmental control and design it should be kept in mind that heat production of chickens can vary as much as 15 to 25% for different reasons.

These data are for floor managed birds. The division of sensible and latent heat production for different environmental temperatures is given in Figure elsewhere in this chapter.

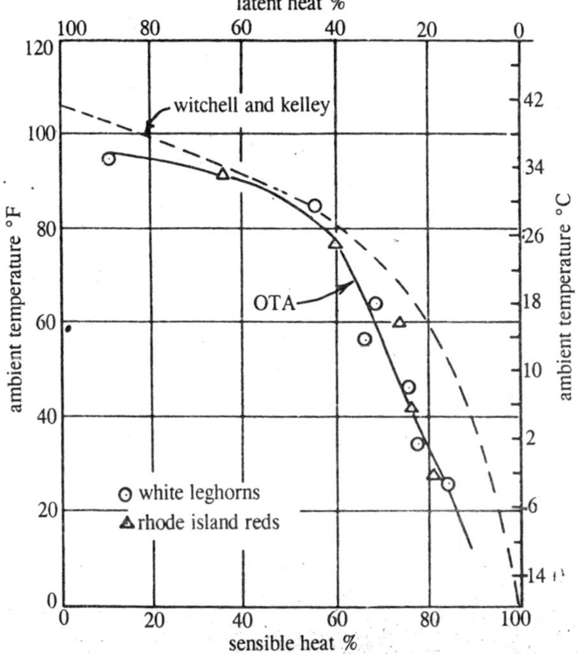

Figure 3.4: Partititon of sensible and latent heat dissipated by laying hens at various envionmental temperatures.

Additional research on the heat loss of caged laying hens is summarized in Figure elsewhere in this chapter. A difference between day and night heat loss will be noted. For White Leghorn hens it was in the range of a 20% reduction during the night at 55°F (12.8°C). The amount of heat reduction was about evenly split for sensible and latent. On unit weight and unit surface area bases, Single Comb White Leghorns lost the greatest amount of heat (latent plus sensible) and the New Hampshire × Cornish Cross the least. The heat loss of Rhode Island Reds was between the other two breeds.

The analysis shown by Figure elsewhere in this chapter indicates that heat loss on a surface area basis tended to reduce breed differences more than the weight basis of Figure elsewhere in this chapter. The heat data directly related to unit body surface area were calculated as follows

$$A = 9.S5\ W^{0.67} \tag{3.4}$$

where A = area in sq. cm.

W = gm of live weight

Feed and Water Requirements

Feed and water input for laying hens may be estimated from Figure elsewhere in this chapter. Feed requirements for poultry are directly related to bird weight, ambient temperature, and rate of egg production. Water input and output is related to feed consumption and environmental temperature.

Part of the water input is from the feed as 54% (wet basis) of the weight of feed is released as water during digestion. This assumes an ordinary moisture content of 10%.

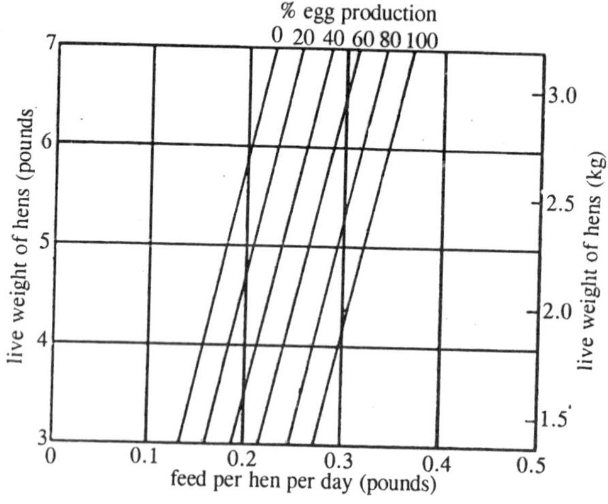

Figure 3.5: Daily feed requirements of hens at various weights and levels of egg production.

As an example a 4.5-lb White Leghorn hen at 75% egg production will require 0.27 lb of feed per day and will consume near twice that much additional water at 55 °F. Water input then is 0.54 lb (2 × 0.27)

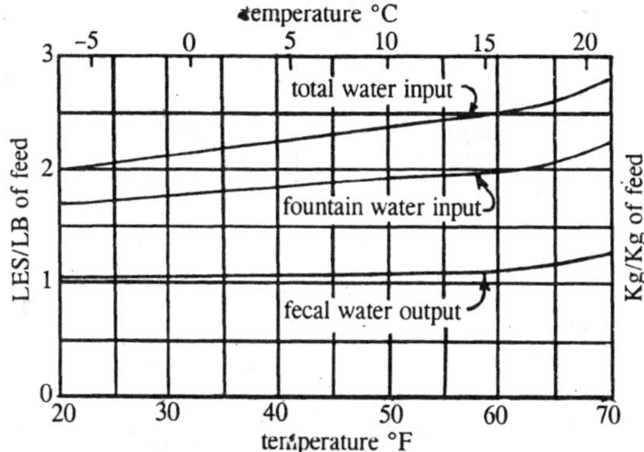

Figure 3.6 : Water input and output per pound of feed cosumed for white leghorn laying hens.

of fountain water plus 54°7o of feed weight (0.27 lb) which is 0.15 lb making a total-of 0.69 lb.

Water is dissipated from the layer in three different ways:

(1) evaporative losses including respiration,

(2) water in feces, and

(3) water in eggs.

Evaporative losses may be estimated from Figures elsewhere in this chapter. Fecal water may be calculated from Figure elsewhere in this chapter, and egg water may be estimated from Figure elsewhere in this chapter.

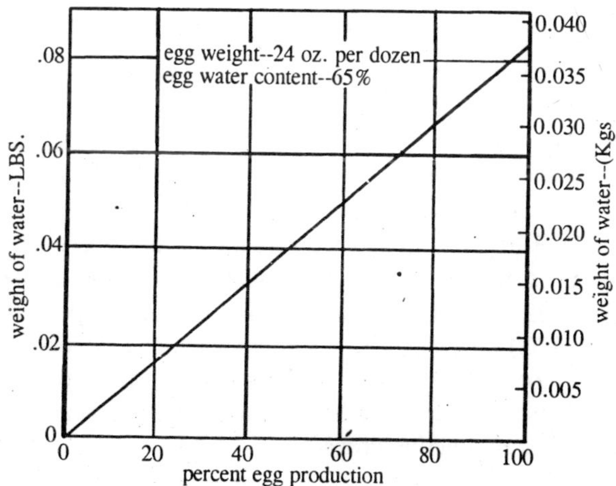

Figure 3.7: Weight of water output in eggs per day per hen at various rates of egg production.

Heat and Moisture Production of Broilers

The total and latent heat loss per pound of body weight of broilers decreases throughout its growth as shown by Figures elsewhere in this chapter. The most rapid change in rate of heat loss is while the broilers are young and light weight. The four different ambient temperatures plotted in Figures elsewhere in this chapter show progressively lower rates of heat loss as the temperature increases.

Moisture production by growing broilers depends on the weight (age) and the environmental temperature. The total respired and fecal moisture produced is needed for ventilation system design. Fecal matter, as defecated, contains 80% moisture. Fecal moisture may be calculated from Figure elsewhere in this chapter and respired moisture from figures elsewhere in this chapter.

Feed consumption of growing broilers at the approximate live body weights and at various ages in weeks is given in Figure elsewhere in this chapter. These data may be used for bin capacity and storage design. However, selected broiler strains will grow much faster and with less feed consumption than indicated in Figures elsewhere in this chapter.

Water consumption is shown by Figure elsewhere in this chapter for growing broilers at an assumed ambient air temperature of 70°F (21.2°C). Growing broilers at 95 °F will consume about twice this amount.

ENVIRONMENTAL REQUIREMENTS

The optimum housing environment for White Leghorn chickens is shown graphically in Figure elsewhere in this chapter. The upper and lower optimum housing temperatures of 85 ° and 55 °F (29.4° and 12.8 °C) provide the desirable temperature range for summer to winter housing.

The 85 °F upper optimum temperature is, however, too high if constant and associated with high humidities. Daytime temperatures may be 85'F (29.4°C) on a diurnal basis if nighttime temperatures drop to 70°F (21.1°C) or lower. The 55°F temperature is a good optimum constant one for winter housing of laying hens. At a constant 45°F (7.2°C) or lower temperature, feed consumption and heat production increase and egg production drops off.

Pound for pound, chickens require three times as much air to supply oxygen as do mammals. Thus for each 24 hr per each 1,000 lb weight

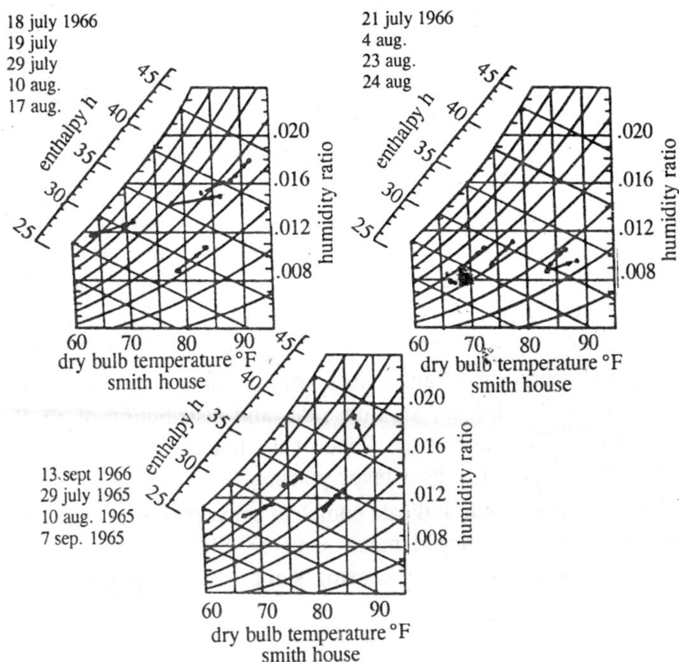

Figure 3.8 : Psychrometric changes in ventilation air as it moves through closed, windowless, caged-laying poultry huses during summer climatic conditions in michigan.

the requirements are:

Man	2,800 cu ft
Cow	2,800 cu ft
Chicken	8,000 cu ft

The health of chickens is closely related to the amount of CO_2 in the air. It should not exceed 900 parts per million (ppm).

Ammonia can be detected by humans at a concentration of near 10 to 15 ppm, and near 50 ppm causes the eyes to water. Research indicates that 40 to 50 ppm of *ammonia* had some effect on *chickens*. A good *ventilation* system should have no trouble keeping the ammonia level at less than 40 ppm.

Minimum *ventilation* air exchange is necessary to maintain air purity, supply oxygen, and remove noxious gases and *airborne* materials. Maximum air flow possible must be maintained during cold weather to remove excess water vapor to keep below 80% RH.

The amount of air flow possible is dependent upon the amount of heat available from the laying hens, or in the case of young birds the *economics* of supplying additional fuel for artificial heat. The summer air exchange must be maximized for maintenance of the inside temperature as little above outside temperature as possible.

The conversion of most or all of the sensible heat of the birds through evaporation of fecal water minimizes the temperature increase in the summer house.

Large pad-type evaporative coolers may help in some climates by adiabatically cooling the air to nearer the wet-bulb temperature. The psychrometric change of summer ventilation air exchange (without an evaporative cooler) in 3 poultry houses is shown by Figure elsewhere in this chapter.

The psychrometric change lines vary in slope and length with outside climatic conditions, air flow rate, and distribution of air flow in the building. The *steeper* the slope of the line the more conversion of sensible heat to latent heat takes place.

A verticle line would indicate complete conversion and one sloping to the left of vertical would mean some sensible heat of the air is converted to latent heat, thus causing some cooling of the dry-bulb temperature.

The intensity of light may be varied for laying hens from ½ ft-candle to about 38 ft-candles with no significant effect on egg production. The number of light and dark hours appears to have more effect than intensity on laying hens. For laying hens, the most common daily lighting program is 14 hr on and 10 hr off.

4

Heat and Water Vapor Production in Farm Animals

Livestock and poultry produce various quantities of heat and moisture. A portion of the moisture is emitted from the animals and birds in the form of water vapor. Some additional moisture is evaporated from moist litter and wet surfaces in the building. For design and operation of an effective environmental control system, the quantities of heat and water vapor produced in the building must be accurately predictable.

For analysis of the steady-state system, discussed in other chapter of this book and expressed by the energy balance equation, the *magnitude* of heat produced by animals or birds q_a and heat added *supplementally* q_s must be calculated. This chapter pertains to these building inputs of heat and water vapor which must be removed by the environmental control system.

HEAT AND VAPOR PRODUCTION

Data are presented on sensible and latent heat production of animals and birds in previous chapters. Some of these data may be applicable for estimating environmental control requirements. More specifically the heat and water vapor production data for the building system when exposed to various *climatic conditions* are necessary for design analysis.

During cold weather seasons the animal heat may be used to warm and *maintain* the building at an optimum temperature. The water vapor produced is the problem in winter. The cold weather air exchange system must remove the excess moisture in the form of vapor to prevent an undesirably high relative *humidity* in the building. If this can be done and an optimum temperature maintained without supplemental heat, maximum *economy* is attained.

During hot weather seasons the sensible heat production of animals and birds is the problem. It must be removed from the building without excessively increasing the building environmental temperature. The moisture and vapor production may, however, be used advantageously to convert sensible heat to latent heat.

Evaporative cooling equipment may be used to maximize this process. Thus, the dry-bulb temperature of incoming air may be lowered to near the wet-bulb temperature.

Heat and Vapor Production in Swine Housing

Building heat losses established from a sow and litter for the first 8 wk after farrowing are shown in Figure elsewhere in this chapter. The division of latent and sensible heat is shown in Figure elsewhere in this chapter along with the progressive weights of the litter, and sow and litter, during the first 8-wk period.

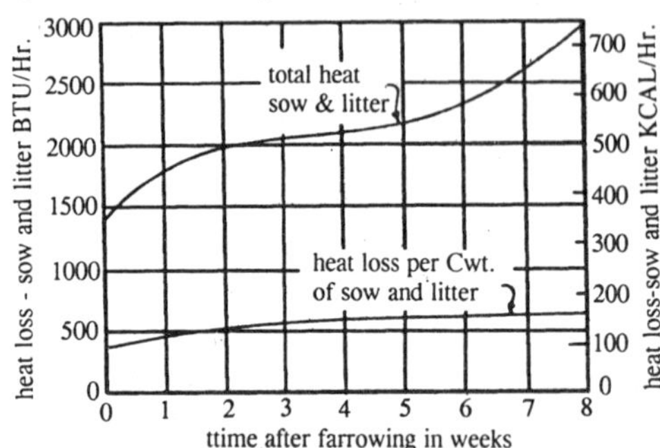

Figure 4.1 : Total heat loss of sow and litter during the first week.

The average litter size for these data was 6 pigs and the temperature range in the farrowing house was maintained from 60° to 80°F (15.6° to 26.7°C). It will be noted that the total heat per hundred weight of sow and litter increased to about 600 Btu/hr within 3 wk and stayed fairly constant through the rest of the 8-wk period.

The division of sensible and latent heat production of the *farrowing* house was nearly even for the first 7 wk. Going into the 8th wk the sensible heat portion became larger than the latent heat.

Total heat production of *growing*, *fattening*, and *mature* swine of

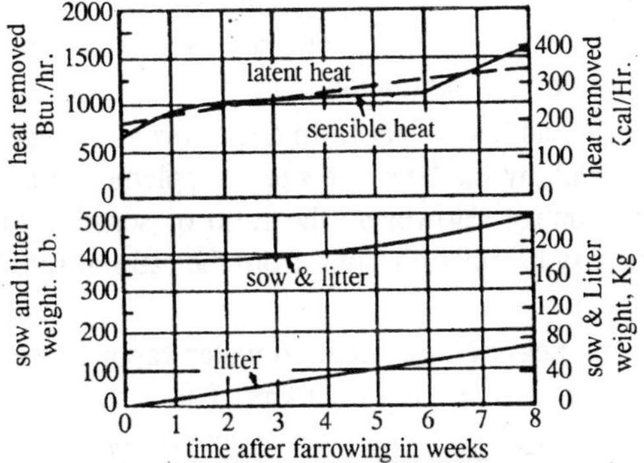

Figure 4.2 : Prtitioned sensible and latent heat ventilated from the farrowing house.

different weights and at different environmental temperatures is given in figure elsewhere in this chapter. The portion of the total that is latent heat is shown in figure elsewhere in this chapter.

These data for heat and water vapor production of hog houses were obtained in a Chamber having a solid concrete floor scraped clear twice

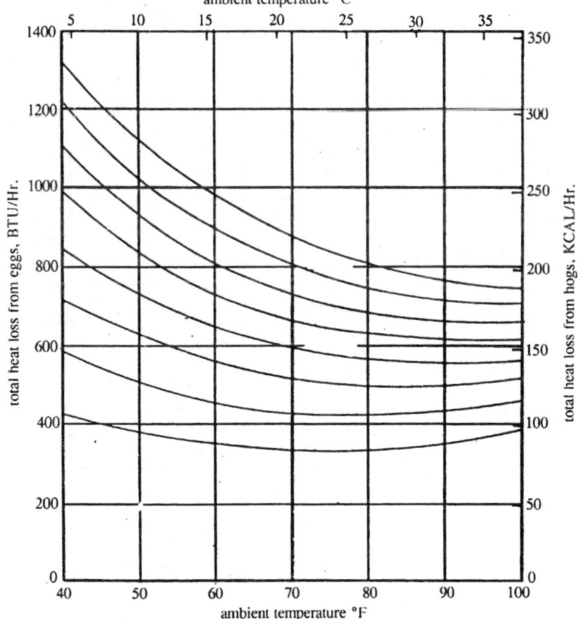

Figure 4.3: Total heat loss for crowing fattening and mature swine.

daily, and no bedding was used. The latent heat loss from the house may vary considerably based upon management practices, building design, and effectiveness of the ventilation system.

The sensible and latent heat from the *calorimeter* type room is the total heat produced by the hogs, plus any supplemental heat added for environmental control. However, the ratio of sensible to latent heat from the hogs may be quite different from the ratio of sensible to latent heat from the room.

Researchers measured the required water vapor removal rate from hog houses with different types of floors. They established significant differences between houses with solid concrete floors and with slotted floors over lagoons. The differences are shown by figures elsewhere in this chapter.

The water vapor produced in a slotted floor hog house (with *lagoon underneath*) is only 0.42 of that in a concrete floor house. In partially slotted floor houses the moisture removal rate is indirectly related to the percent of the floor that is slotted. The data for water vapor removal from a concrete floor hog house may be expressed *mathematically* by the following regression equation

$$Y = -0.961 + 0.291x_1 - 0.78.5x_2 - 0.146x_1x_2 - 0.029x_1^2 + 1.375x_2^2 \quad (4.\ 1)$$

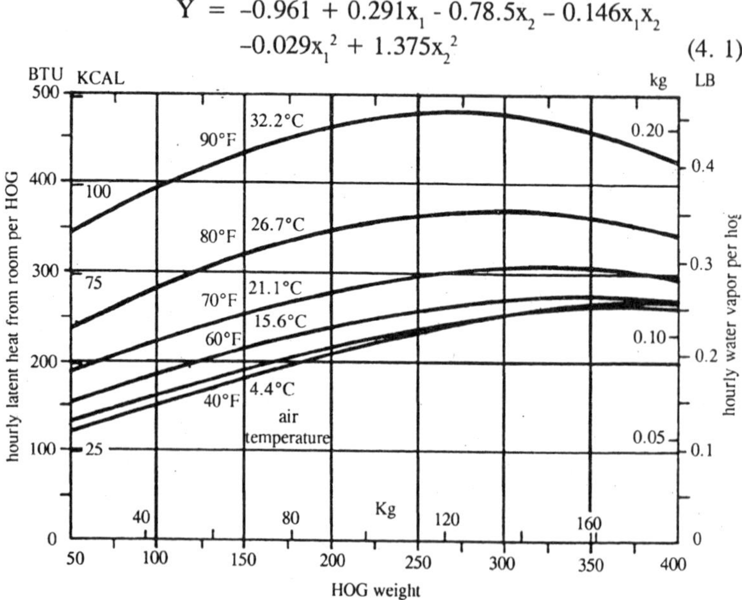

Figure 4.4: Total room latent heat from a hoc house with solid concrete floors.

where Y = logarithm to base 10 of the moisture removal rate in lb/(hr) (hog)

x_1 = body weight/100, lb

x_2 = air temperature/100, °F

The effect of temperature on water use by swine is shown by figure

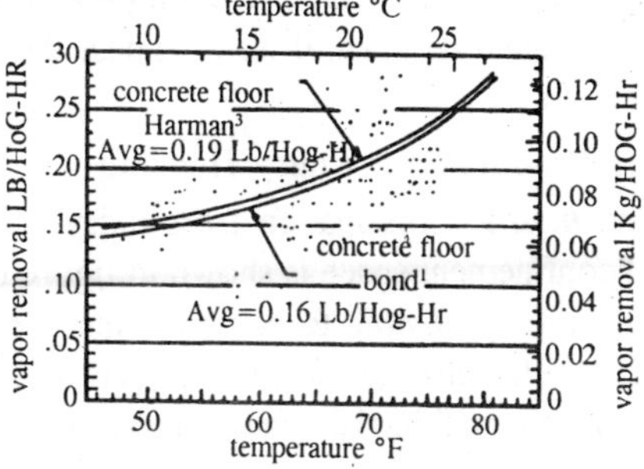

Figure 4.5: Water vapor removal rates from swine houses with concrete floored pens.

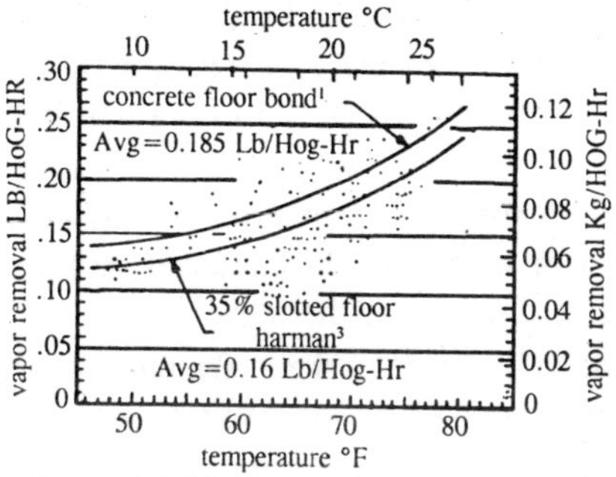

Figure 4.6: Water vapor removal rates from swine houses with partially slatted floors.

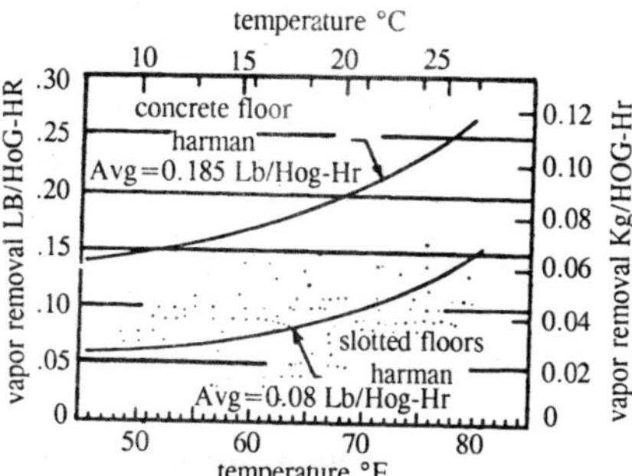

Figure 4.7: Water vapor removal rates from swine houses with totally slatted floors.

elsewhere in this chapter. These data indicate that about twice as much water vapor is removed from the building, with relative *humidity* at 50%, than the hogs produce directly m the form of vapor.

As might be predicted, all lines of figure elsewhere in this chapter tend to converge at an environmental temperature of 100°F. Nearly all water drunk is then *eliminated* from the building as water vapor. Thus at high temperatures the total input of water is near the amount removed by the ventilation air.

Heat and Vapor Production in Dairy Housing

The heat and *moisture* dissipation from dairy barns as related to temperature and confinement space is shown in figure elsewhere in this chapter. The data are *applicable* to cows in *stanchions* under typical barn conditions. No allowance for heat from lights and equipment or for heat emitted by people within the space has been made.

The moisture curve of figure elsewhere in this chapter does include moisture evaporated from *stall*, *gutter*, and *manger* surfaces. Thus, it is basically the moisture that must be handled by the *ventilation* system. The data apply to both Jersey and *Holstein* breeds, although the values may be from 5 to 10% low for Jerseys on a unit body-weight basis.

For purposes of comparison the mositure-vapor curve of figure elsewhere in this chapter may be read in pounds of water or Btu's of latent heat. The latent heat of vaporization was assumed to be 1,044 Btu per lb. The change in shape of the mois-

Lure curve at 65°F (18.3°C) shows the point where physiologically the cow increases her rate of evaporation from lungs and body surface.

Dairy housing is not always maintained at a constant temperature and most natural *climatic* conditions have some diurnal variation. Studies were made at the University of Missouri to determine heat and moisture production (as well as other factors) of dairy structures under diurnally varying environmental housing conditions.

Figure elsewhere in this chapter shows the average daily rate of stable heat dissipation, and figure elsewhere in this chapter the average daily rate of stable *moisture* production during different diurnal temperature variations. The results compare fairly well with constant temperature data, when the diurnal average temperature is used as a base.

Theoretically the heat dissipation should be lower under *diurnal* conditions, because heat production drops at an increasing rate with rising temperatures. The depressing effect of temperatures above the

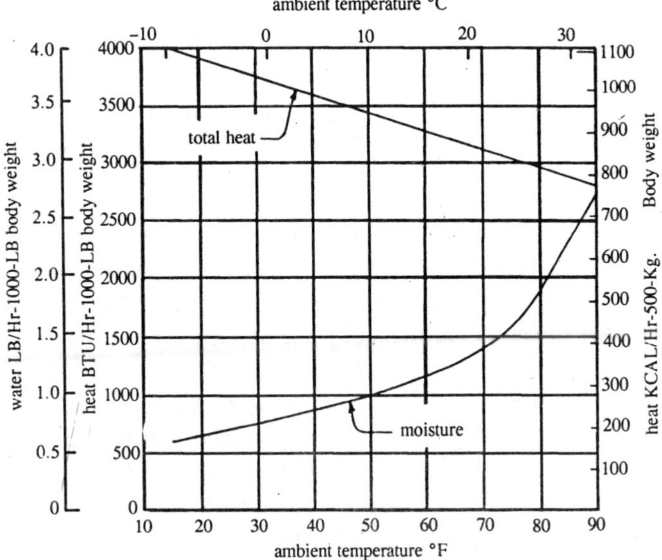

Figure 4.8: Total heat and water vapor dissipation rates for stanchioned dairy cows.

average temperature would therefore overbalance the elevating effect of the lower temperatures. This was true only during the 40° to 70 °F (4.4° to 21.1°C) and 70° to 100°F (21.1° to 37.8°C) diurnal tests.

Diurnal changes in stable heat and moisture dissipation with diurnal ;table temperature changes are shown in figure. It will be noted that these heat dissipation curves neglect storage. A negative heat dissipation value is in evidence on the 60° to 110°F (15.6° to 43.3°C) diurnal curve. This in actuality does not reflect metabolic heat production directly as it never dropped below 2,700 Btu/ (hr) (cow).

The diurnal changes in stable moisture *dissipation* shown in figure elsewhere in this chapter indicate a 2 to 4 hr lag when compared with maximum and minimum temperatures. The magnitude of the diurnal moisture variation was least (about 0.14 lb/ (hr) (cow) for the 10° to 40°F (-12.2° to 4.4°C) range, and greatest (about 1.5 lb/ (hr) (cow) for the 60° to 110 °F (15.6° to 43.3 °C) range.

High *diurnally* varying stable temperatures will necessitate the consideration of diurnal moisture dissipation in ventilation system design. The 3.05 lbi (hr) (cow) (1.38 kg/ (hr) (cow)) water *dissipation* rate following the 100°F (37.8 °C) temperature of the 70° to 100°F (21.1° to 37.8°C) diurnal is 35°79 above the average rate of 2.25 lb/ (hr) (cow) [(1.04 kg/ (hr) (cow)].

Heat and Vapor Production of Beef Cattle

The optimum environmental temperature for beef cattle varies with breed, age, weight, and condition. The heat tolerance difference between European and Indian cattle, up to 95°F (35°C), is about 20°F (11.1°C). The "*comfort zone*" (temperature interval during which no demands are made on the temperature *regulating* mechanisms) of European cattle is between 20° and 60°F (-6.7° to 15.6°C), and for Indian cattle between 50° and 80°F (10° and 26.7°C).

Total heat production is influenced by thermal environment, animal breed, condition, age, weight, and *plane of nutrition*. Figure elsewhere in this chapter shows the effect of plane of nutrition on heat production at a constant temperature of about 64°F (17.8°C) (curves A and B).

The plane of nutrition (shown on the top x-axis of graph) range from fasting to three times maintenance ration. The heat production more than doubled with this variation of ration. At the "*maintenance*" level, heat production increased by about 17% when hay alone was fed.

Studies related animal weight (lower x-axis scale) to heat production at two constant environmental temperatures of 50° and 80°F (10° and 26.7°C). These animals were fed corn and hay at a level of from 2 to 2.5 times the maintenance ration previously used.

Factors affecting the ratio of latent heat to total heat are: age, weight, condition and breed of animal, mean radiant temperature, ambient temperature, and plane of nutrition. Figure elsewhere in this chapter shows the effect of environmental temperature on the percentage of latent heat for 3 breeds—*Shorthorn*, *Brahman*, and *Santa Gertrudis*—at ambient temperatures between 65° and 106°F (18.3° and 41.1°C). It is to be noted that at 100°F (37.8°C) and above the amount of heat loss as vapor is greater than the heat produced.

This indicates that the animal is actually gaining heat by other methods. The lower curve B of figure elsewhere in this chapter relates percent latent heat to plane of nutrition with a maintenance ration of ground corn and alfalfa hay. The curves correspond quite well when it is considered that the nutrition tests were at a constant calorimeter temperature of 64°F (17.8°C) and the temperature tests were carried out on rations equivalent to about twice maintenance level.

Surface temperatures of animals largely determine heat loss by radiation and convection. They may also be an indication of comfort or tolerance to temperature extremes, both hot and cold. The surface temperature data of figure elsewhere in this chapter show skin and hair temperatures increasing with environmental temperatures for Shorthorn,

Brahman, and Santa Gertrudis calves of 750 to 900 lb. Also given in the lower and right portion of figure elsewhere in this chapter is the relationship between animal live weight and surface area.

Research at the University of Missouri showed that at 80°F (26.7°C) constant temperature, increasing body weight was associated with decreasing skin temperature and increasing hair temperature. At 50°F (10°C), both skin and hair temperature decreased with increasing weight.

HEAT PRODUCTION OF SHEEP

Heat production data on sheep as affected by the energy level of feeds, environment temperature, and weight of sheep are given in figure elsewhere in this chapter. The 3 curves representing different energy levels of feeds of figure elsewhere in this chapter were for sheep with closely *clipped fleeces* averaging only about 0.1 cm (0.039 in.) in length.

Studies were made on the effect of fleece length upon heat production by sheep at environmental temperature of 68°F (20 °C). Heat production decreased rapidly and *exponentially* from a maximum of 380 Btu/hr (95.8 kcal/hr) with a closely clipped fleece to a minimum of about 24 Btu/hr when the fleece had grown to a length of 20 to 30 mm (0.79 to 1.18 in.).

When the temperature was decreased to 46°F (7.8°C), the range in heat production was from about 520 Btu/hr (131 kcal/hr) with closely clipped fleece to 265 Btu/hr (66.8 kcal/hr), when the fleece had grown to 45 mm (1.77 in.). These studies were at the medium level feed energy.

Surface area of unshorn sheep may be determined by the formula

$$A = 0.124\ W^{0.561} \tag{4.2}$$

where A = surface area in m^2

NN' = live body weight in kg

or $A = 0.856\ W^{0.561}$

where A = surface area in ft^2

W = live body weight in lb

Priestly describes the heat balance of a sheep as

$$Q_m = q_r + q_c + q_e + q_s$$

where Q_m = rate of heat production in body

q_r = net rate of radiation loss

q_c = rate of conduction and convection loss

q_e = rate of evaporative heat loss

q_s = rate of heat storage

It has keen estimated that a standard unshorn sheep, considered as a horizontal *cylinder* 1.0 m (39.37 in.) and 0.5 m (19.685 in.) in diameter, will experience under high sun a total receipt from shortwave radiation as high as 500 kcal/hr (1,985 Btu/hr).

This is many times the metabolic heat production rate and many times the rate at which a sheep is able to *compensate* for through physiological mechanisms of the heat loss. The compensations must therefore be physical processes which are automatic and take place externally to the sheep.

It is more *conservatively* estimated that direct shortwave radiation might amount to 250 kcal/hr (993 Btu/hr) and diffuse radiation from the sky 60 kcal/hr (238 Btu/hr) with another 45 kcal/hr (179 Btu/ hr) reflected from the ground for a total of over 350 kcal/hr (1390 Btu/ hr),

The absorption of shortwave radiation by the wool tips has been reported to raise tip temperatures to as high as 190°F (87.8°C) and temperature difference of 45°C (81°F) across 4 cm (1.58 in.) of fleece. This would be the maximum temperature difference high on the animal back.

If half of this difference is taken as an average over the body of the sheep, a conduction rate to the body through the fleece of about 32 kcal/hr (131 Btu/hr) is implied. This is comparable to both the *metabolic* rate and evaporative heat loss. The effective thermal conductivity of wool is taken throughout at 1 × 10° kcal/ (sec) (cm) (m2) (°C), [0.1875 Btu/ (hr) (in.) (ft^2) °F].

The conduction rate to the body thus is only about 1/10 of the radiation load. This is very significant with respect to body temperature regulation as most of the heat must be dissipated from the fleece by convection and longwave radiation. Both means of heat transfer, however, increase with increased temperatures of the wool tips over that of the environment.

Two equations of heat balance may be written: one of the tips and another of the body of the sheep. The first is predominantly under physical control and might be termed an external heat balance equation

$$q_{rr} = q_{re} = q_{cf} + q_{ca} \tag{4.4}$$

where Q_{rr} = net receipt of shortwave radiation

q_{re} = net loss of longwave radiation

q_{cf} = conduction inward through fleece

q_{ca} = convective losses from fleece to air

The second is an internal heat balance which is under physiological control and is the metabolic heat production equivalent

$$Q_m = q_{rs} - q_{cf} + q_{er} + q_{es} + q_s \qquad (12.5)$$

where q_{rs} = respiratory sensible heat loss

q_{cf} = conduction inward through fleece

q_{er} = evaporation heat loss by respiration

q_{es} = evaporation heat loss from skin

q_s = body heat storage

The ultimate concern insofar as the animal's ability to maintain *homeothermy* is equation. The significant factor here is q_{cf}, the purely physically controlled heat conduction gain to the animal through the fleece.

This factor is governed by the external climatic conditions which directly impose stresses on the sheep. The problem then is how to determine q_{cf} which is a residue quantity of heat between large factors not accurately known.

Heat Loss by Convection

Heat loss by convection q_{ca} depends primarily on the dimensions of the sheep (regarded as a horizontal cylinder), on the wind speed v taken at right angles to the *cylinder axis*, and on the average difference of temperature At between the wool tips and the surrounding environmental air. Assuming forced convection, the heat loss depends on the Reynolds number and the temperature difference.

Under calm condition, air buoyancy must be considered. Calculations on fleece convective losses are plotted in figure elsewhere in this chapter. The data are based on experiments with cylinders of uniform surface temperature which does not truly hold for sheep. For animals of different size the convective heat loss per animal will vary with animal (cylinder) length (1) and diameter ($d^{0.7}$) at the strongest winds; and as $(1d)^{0.9}$ under the calm conditions.

Longwave Radiation Loss q_{re}

For a sheep with specific geometric dimensions of length 1 and diameter d the longwave radiation heat loss depends mainly on air temperature t_s, ground temperature t_g, surface temperature of the hair tips t_s, and vapor pressure p, of the *atmosphere*. An approximation of the radiation losses may be written

$$q_r = 1/2\pi d1(2\sigma T_s^4 - \sigma T_g^4 - 1.04B) \qquad (12.5)$$

where B = the intensity of back radiation from the atmosphere received per unit area of horizontal surface

σ = Stefan-Boltzmann constant

An estimation of B is as follows

$$B = (0.44 + 0.08\sqrt{p_W})\,\sigma T_a^4 \qquad (12.6)$$

where p_W = atmosphere vapor pressure in millibars

Equation requires a ground temperature to represent longwave radiation from the ground incident upon a horizontal surface. It may be measured by a downward-pointing radiometer. This would provide measurements for each specific case. A more general approach is to assume $T_s = T_g$ thus

$$q_r = 1/2\pi d1(\sigma T_s^4 - 1.04B) \qquad (12.7)$$

Table elsewhere in this chapter presents some results of such calculations with d = 50 cm

Table 4.1: Longwave radiation loss from standard sheep (Kcal/hr)

t_a			90°F					110°F			
t_s			100	130	140	160	180	120	140	160	180
Vapor Pressure											
Mm Hg	*Psi*	*Mb*			*kcal/Hr*				*Kcal/Hr*		
3.75	0.0725	5	148	202	260	327	401	171	229	296	370
7.50	0.1450	10	123	179	237	303	378	140	198	265	339
15.0	0.2900	20	82	136	195	261	335	97	155	222	296
					Btu Hr				*Btu/Hr*		
3.75	0.072	5	587	800	1030	1297	1588	678	908	1175	1467
7.50	0.1450	10	488	710	938	1198	1496	554	786	1050	1342
15.0	0.2900	20	325	538	773	1033	1326	384	614	980	1172

(1.64 ft), 1 = 100 cm (3.28 ft), t_a =90° and 110°F and a range of p_w and t_s values.

Heat Conduction Through the Fleece

This is the other unknown for solving equations (4.4) and (4.5). It is assumed that shortwave radiation intensity, environmental conditions of air temperature, vapor pressure, and wind are known, as well as approximate body temperature of the sheep, and the length and thermal conductivity of *fleece*.

Surface temperature of the wool tips is the main unknown. The surface temperature t_s and conduction heat transfer of fleece q_{cf} may be found by a series of approximations. The q_{cf} may first be assumed = 0 in equation (4.4) and a t_s, obtained. From this Δt a (q_{cf}) may be calculated and substituted in equation (4.4) etc.

5

Air Exchange in Farm Animals

The remaining factor of the heat balance equation to be discussed in this chapter pertains to heat transfer through air exchange. The heat input and output for a *livestock* or poultry building must be equal, under steady-state environmental conditions. A predictable amount of sensible heat will be *transferred* through the walls and exposed surfaces of the building.

The remaining output of heat must be accounted for through air exchange. Air exchange is the *manipulatable* portion of the livestock *housing system* and in fact must continually be controlled.

The amount of conduction heat loss through the exposed surfaces can be manipulated by varying the amount of insulation, shape of the building, and inside temperature. Once the building has been designed and constructed, however, the first two items are *fairly* well fixed.

Insulation and heat transfer are discussed in other chapter of this book. Figure elsewhere in this chapter shows the effect of building shape on exposed area. It is noted that size and width of building are not very significant from the standpoint of heat loss.

Inside temperature depends upon the *optimum* requirement for the livestock and/or poultry housed. It should be established according to these requirements and not manipulated appreciably thereafter.

Heat input by the animals or birds is also pretty well fixed in a building depending on management decisions pertaining to number, size, production, and nutrition level. The amount of *supplemental* heat provided

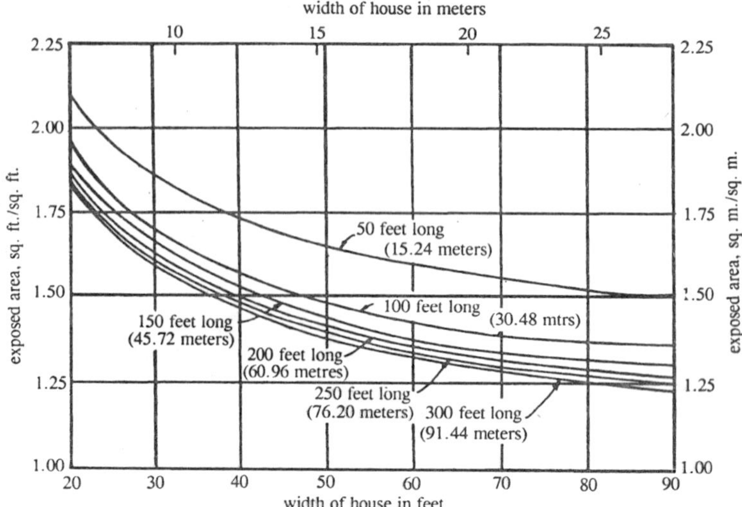

Figure 5.1: Exposed surface area in sq. ft. per sq. ft. of floor area for livestock and poultry house with 8-ft. ceiling height.

from artificial fuels or electricity is variable. The amount of heat added, if any, during cold weather depends upon the outside temperature, the *environmental* temperature, and *ventilation needs* of the animals or birds and the overall economics of housing.

In the northern part of the United States, supplementary heat is normally required for young animals and birds. Thus, a means of providing supplementary heat is necessary for *farrowing houses*, *lambing buildings*, *calf houses*, and *growing* or *brooder* houses for chickens.

Housing for adult animals and birds in the northern part of the United States does not normally require *supplemental* heat if the building is designed properly and the air exchange controlled effectively.

HEAT BALANCE CALCULATIONS FOR COLD WEATHER OPERATIONS

The following steps are suggested. (1) Decide on number and size of animals or birds for the building (management decision). (2) Calculate total heat input. (3) Calculate the average heat transfer coefficent U_{av} for the building. (4) Calculate exposure factor for building. (5) *Determine* inside design temperature (*management decision*). (6) Determine outside design temperature. (7) Calculate heat available for warming ventilation air exchange and vaporizing moisture. (8) Determine inside and outside design relative humidities or the proportion of total heat to be removed from the building as latent heat.

Heat Loss Per Animal or Bird

The number of animals or birds is variable, but *independent* from the standpoint of the engineer (it may be varied some by management within a *specific building*). The conductive heat loss may be calculated on a per animal or per bird basis once the number is established. This term is called an exposure factor and may be expressed mathematically as

$$\text{E. F.} = \frac{A_t U_{av}}{N} \quad (5.1)$$

where E.F. = exposure factor
A_t = total exposed area of walls and ceiling of building
U_{av} = average overall coefficient of heat transfer for the building
N = number of animals or birds housed

By definition the exposure factor is the conductive heat loss from a building per animal or bird per degree temperature difference. For a given building, the better it is insulated the smaller the *exposure* factor; and the *denser* the population of animals or birds in the building the smaller the exposure factor.

Example 5.1—If a 40 by 200 ft poultry building with an 8-ft ceiling height houses 10,000 laying hens, and the U_a, is 0.15 Btu/ (hr) (ft^2) (°F), what would be the exposure factor?

Area of ceiling = 40 × 200 = 8,000 ft^2
Area of sidewalls = 8 × 480 = 3,840 ft^2
Total area of exposed surfaces = 11,840 ft^2

$$\text{E.F.} = \frac{A_t U_{av}}{N}$$

$$\text{E.F.} = \frac{11{,}840 \times 0.15}{10{,}000}$$

$$\text{E.F.} = 0.178 \text{ Btu/(hr)(°F)(hen)}$$

Exposure factors for laying houses will normally be in the range of from 0. 1 to 1.0 Btu/ (hr) (°F) (hen).

Example 5.2—A 24 by 60 ft farrowing house is designed with 20 farrowing stall units and insulated to have a U_{av} of 0.10 Btu/ (hr) (ft^2) (°F). If an 8-ft ceiling is assumed, what would be its exposure factor?

$$\text{E.F.} = \frac{(1440 + 1344) \times 0.10}{20} = 14 \text{ Btu/(hr)(°F)(sow)}$$

Example 5.3—For a 36 by 100 ft finishing house for 360 hogs, what would the exposure factor be if an 8-ft ceiling is assumed and U_{av} = 0.12 Btu/ (hr) (ft²) (°F)?

$$E.F = \frac{(3600 + 2176) \times 0.12}{360}$$

$$E.F. = 1.93 \text{ Btu/(hr) (°F) (hog)}$$

In example 5.3, it will be noted that 10 ft² of floor area per animal was allowed. Depending upon the size and age of hogs as well as management practices, the floor area per animal might be considerably less than 10 ft² per hog or in some cases more.

The exposure ratio for hog houses (including farrowing as well as finishing) will normally be in the range of 1.0 to 20.0 Btu/ (hr) (°F) (hog).

Example 5.4—A stall-barn for dairy cattle 36 by 120 ft is designed for 54 cows. If the U_{av} = 0.20 Btu/ (hr) (ft²) (°F) and the ceiling height is 8 ft, what is the exposure factor?

$$E.F. = \frac{(4320 + 2496) \times 0.20}{54}$$

$$E.F. = \frac{6816 \times 0.20}{54} = 25.2 \text{ Btu/(hr)(°F)(cow)}$$

The exposure ratios for dairy or beef barns may vary from 10 to 100 Btu/ (hr) (°F) (cow) depending upon building design and management practices.

The exposure factor definition for a building might be altered to make heat loss per square foot of floor area per degree *Fahrenheit* difference the Common *denominator* rather than per animal or bird. This definition might be used if there is a desire to make comparisons between buildings for *hogs*, *cattle*, and *poultry*.

Normally for specific livestock building analysis the per animal or per bird basis is preferred because the heat and vapor input varies directly with the number of animal or bird units. Also ventilation air exchange must be designed for the number of animals or birds and not necessarily just for the size of the building.

Conduction Heat Loss Per Animal or Bird

Total conduction losses may be determined directly from *graphs* designed for that purpose. This step of the heat balance calculation introduces the temperature difference for the building based upon inside and outside deign conditions.

Figure elsewhere in this chapter presents graphically the heat loss for various temperature differences and exposure factors typical for poultry houses. For example a temperature difference of 50'F (which might represent an inside temperature of 50°F and an outside temperature of 0°F) would allow a heat loss of 40 Btu/(hr) (bird) if the exposure factor for the building was 0.8 Btu/(hr) (°F) (bird).

From data in other chapter of this book it will be found that 40 Btu per bird is the total heat production of a typical White Leghorn laying hen at about 4½ lb and 55°F. This means that *theoretically* all of the heat produced by each bird is lost through the exposed surfaces of the building. There would be no heat available for ventilation.

Unless supplemental heat is provided, this would be a very undesirable condition. The E. F. of 0.8 represents a very poorly insulated building and/or also one which is sparsely populated with birds. In any case it should not be constructed in a climate where there is any possibility of a 50°F temperature difference between inside and outside. For cold climates the economics of operation would dictate that the house be better insulated to save on the cost of *supplemental* heat.

For hog housing figure elsewhere in this chapter provides a graphical

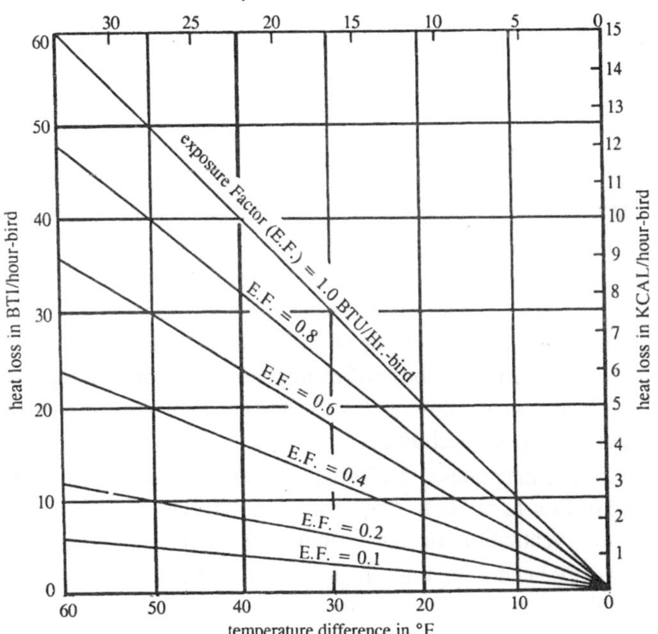

Figure 5.2: Conduction heat loss per bird through walls and ceiling for 1°F temperature difference.

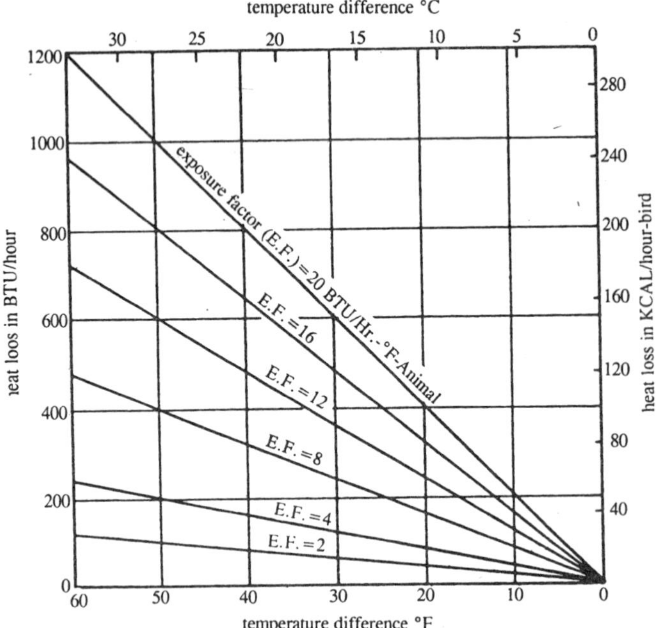

Figure 5.3 : Conduction heat loss per hog through walls and ceiling for 1°F temperature difference.

means of determining the heat loss per hog for a specific temperature difference and exposure factor. For a more typical example than the extreme one assumed for poultry housing, use an exposure ratio of 1.93 for the finishing house of example 5.3.

Assume an inside temperature of 50°F and outside design temperature of 20T (At = 30°F). From figure elsewhere in this chapter the heat loss through the exposed surfaces per hog may be estimated as 60 Btu/(hr) (hog). For the farrowing house of example 5.2 with an exposure ratio of 14 the heat loss would equal 420 Btu/(hr) (sow) (a higher inside temperature might be desirable).

For the finishing house, if 100-lb hogs were assumed, each of which produce 510 Btu/ hr at 50°F, there would be 450 Btu (510-60) available for warming ventilation air and evaporating water.

Figure elsewhere in this chapter may be used to make similar estimates for cattle housing structures.

Possible Ventilation Air Exchange

After deducting the heat loss through the exposed surfaces of the building, the remainder of the heat produced per animal or bird is available for warming ventilation air exchange and for *vaporizing* water. At this point the *psychrometric* chart may be used to estimate the air exchange possible.

The latent heat method or *enthalpy* method may be used. For the first method the portion of heat to be removed in the latent form must be established. The building heat loss charts of other chapter of this book may be used for this.

The remaining available sensible heat may be used for warming incoming ventilation air. This division of latent and sensible heat removed in the ventilation air is difficult to estimate accurately. The latent heat estimate must include evaporation from the litter and moist building surfaces as well as evaporation directly from the animals or birds.

The second method consists of calculating the enthalpy change of the ventilation air as it moves through the building. For this, design levels must be established for inside and outside relative humidity as well as inside and outside temperature.

Fortunately these relative humidities can be quite accurately *predicted* and the variation is small. Observations of winter operation of poultry laying houses indicate that the exhaust air will run in the range of 70 to 80% with an effective ventilation system.

A relative humidity above 80% becomes undesirable from the standpoint of bird health and management; however, if it drops appreciably lower than 70% the effectiveness of the system in removing evaporated water decreases.

Figure elsewhere in this chapter present graphical solutions by the enthalpy method for determining the quantity of ventilation air exchange possible for poultry, swine, and cattle housing respectively. These solutions are for design conditions of 50 °F and 80% RH inside and 80% RH outside. The outside design temperature is the variable on the x-axis of the graphs.

Problems that have other inside design conditions must be solved directly with the *psychrometric* chart. An assumed 80% RH outside for fairly cold climatic conditions can only introduce a small error, because the absolute moisture content of cold air is low even if saturated.

Example 5.5—Determine the air exchange possible per laying hen with an outside design temperature of 20°F, 80% RH outside design, and inside conditions of 50°F and 80% RH. The house is similar to the one described in example 5.1.

From example 5.1 an exposure ratio of 0.178 was determined for this specific poultry house. From figure elsewhere in this chapter the conductive heat loss for Δt = 30°F is estimated as 5.3 Btu/(hr) (bird). Heat production for a 4.5 lb White Leghorn layer at 50°F is estimated at 40.5 Btu/hr.

Heat available for warming ventilation air exchange and evaporating water is (40.5-5.3) approximately 35 Btu. From figure elsewhere in this chapter for an outside temperature of 20°F it may be determined that 35 Btu/hr will provide an air exchange of something over ½ cfm (possibly near 2/3 cfm). This more than meets the specified minimum *ventilation* requirements. If it did not meet the minimum ventilation requirements a better insulated structure or supplemental heating would be required.

The amount of vaporized moisture removed is critcal. For the example 13.5 building it may be determined from figure elsewhere in this chapter. It will be noted that a ventilation rate of ½ cfm at 20°F outside temperature will remove about ¼ lb of water per day per bird.

This is approximately the rate of direct evaporation from a 4.5 lb layer at 50°F. The *additional* ventilation possible for the system of example 5.1 would provide some net drying of the litter or *droppings* during a winter having a design temperature of 20°F. The example 5.3 for a swine finishing house may be followed through in a similar way to determine that air exchange of 8 cfm per hog is possible with the heat available.

Latent Heat Method of Calculating Possible Ventilation Air Exchange

A somewhat modified latent heat method approach was proposed for swine finishing houses. It applies the sensible heat production of the animals directly to warming *ventilation* air exchange. Figure elsewhere in this chapter shows graphically the ventilation air exchange possible with one set of design conditions.

A finishing house for 50- to 100-lb hogs with the inside temperature at 50°F is used for this example. Exposure factors of 1, 2, 3, and 4 are shown with 2 superimposed relative *humidity* curves (80 and 100%). The exposure factor curves are based on a sensible heat production of 310 Btu/(hr) (hog) and the moisture curves on a latent heat production of 146 Btu/(hr) (hog).

In comparison to example 5.3 previously discussed, if figure elsewhere in this chapter is checked to determine possible ventilation at 20°F outside temperature and with an exposure factor of 2, it will be found to be slightly over 7 cfm/hog.

This checks fairly closely with the 8 cfm/hog obtained from figure elsewhere in this chapter. The curves of figure elsewhere in this chapter are based on a lower heat production per animal for hogs between 50 to 100 lb where the result from figure elsewhere in this chapter is based upon 100-lb hogs.

The design of figure elsewhere in this chapter is based upon constant latent heat or moisture removal from the building regardless of building exposure factor, amount of ventilation or outside design temperature.

The moisture removal is determined by the latent heat produced directly by the animal through respiration and evaporation from body surfaces; this excludes any evaporation from litter, feces or wet surfaces. The previously discussed enthalpy method for determining possible ventilation air exchange is based upon a constant 80% RH.

This is also impossible to hold constant under actual operating conditions. Field studies indicate, however, that thermostatically controlled systems will hold close to this even with varying outside winter temperatures. The 80% RH should be considered a maximum relative humidity.

When there is not enough ventilation air exchange possible to remove the latent heat produced directly by the animals or birds, more heat is necessary and/or more *insulation*. The additional heat might be supplemental heat or from a denser population of birds or animals housed in the building.

The latent heat approach has been taken one step further to show possible ventilation air exchange control. Figure elsewhere in this chapter indicates the requirements for a *specific* swine building with an exposure factor of 1.1.

This method is based upon a constant minimum rate of air exchange up to 45°F outside temperature and a constant maximum 80% RH throughout the same range. A varying inside temperature of from 50° to 70°F is thus a built-in result of this method of control.

Fan control is simplified, however, as only one rate of air delivery for the building would be necessary for normal winter temperature conditions. Figure elsewhere in this chapter is based upon a constant ventilation air exchange rate of 6 cfm for each 50- to 100-lb hog.

Below 10°F outside temperature it would not be possible to hold a minimum of 50°F inside and a maximum of 80% RH; thus supplemental heat would be necessary.

HOT WEATHER ENVIRONMENTAL CONTROL

The trend of poultry housing in the northern part of the United States is for *closed*, *insulated*, *windowless* buildings. Mechanical ventilation must therefore be provided continuously during hot weather as well as cold weather. The trend of other livestock housing seems also to be towards *windowless* environmentally controlled structures.

Various means of cooling the enclosed environment will also be introduced and practiced in the future as economics justify. High temperatures in the United States are more harmful to livestock and poultry production than low temperatures. The excess heat is also more difficult to cope with from the standpoint of environmental control. Figure elsewhere in this chapter illustrates the detrimental effects for poultry egg production.

The low environmental temperatures can be economically offset by utilizing the heat from the birds and animals. In hot weather the excess heat is an extra burden and mechanical cooling is costly to install and operate.

The hot weather heat production of the animals or birds must be removed from the building in either the sensible or latent form. Fortunately the total heat production of animals and birds is less at higher environmental temperatures. Also, a greater portion of the total is dissipated directly by the animals or birds in the latent form.

The remaining sensible heat produced must either be absorbed by the ventilation air in the sensible form or converted to latent heat. Sensible heat absorption causes a temperature increase of the ventilation air exchange and ambient conditions in the house.

A study of sensible heat conversion to latent heat in cage-type laying houses found that some houses consistently exhausted less than 25% of the total heat production as sensible heat while others *fluctuated* up to 50% and above.

The amount of this effective evaporative cooling within the poultry house depends upon: (1) the amount of ventilation air exchange through the house (4 cfm/bird is minimum, 5 or 6 cfm is optimum); (2) distribution of the ventilation air throughout the house and the exposure of moist droppings (the ideal is well coned droppings below the cages); and (3) low vapor pressure or *relative humidity* of incoming ventilation air. High outside temperature (90°F or above) conditions appeared to bring about higher ratios of latent heat removal even when the relative humidity was fairly high (50% or higher).

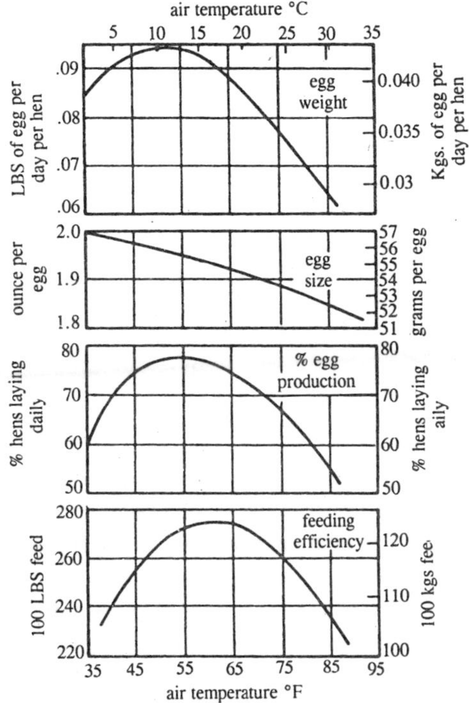

Figure 5.4: The effect of temperature on egg weight, egg size, egg production, and feed efficiency for hens.

In only I house on 1 specific day, out of 6 houses studied during some 25 different summer days, was there found to be throughout the day a lower dry-bulb temperature inside the house than outside. Thus, on this particular day all of the heat from the birds was being converted to latent heat and some additional evaporation was utilizing some of the sensible heat of the incoming ventilation air.

This of course is a very desirable situation but not often attained without evaporative cooling equipment.

Figure elsehwhere in this chapter shows graphically the possible psychrometric change of the ventilation air as it passes through a livestock or poultry building. Point O represents on a psychrometric chart the temperature and relative humidity of the incoming ventilation air (for example 90°F dry-bulb and 40% RH).

The direction and length of the line from point O then indicates the drybulb air (for example 90°F dry-bulb and 40% RH). The direction and length of the line from point O then indicates the dry-bulb temperature

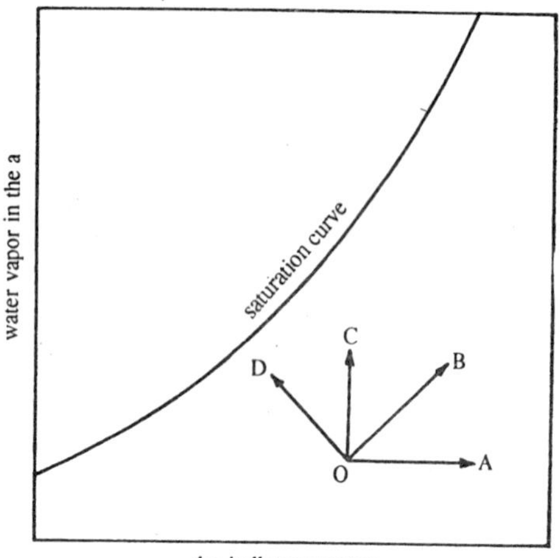

Figure 5.5: Psychrometric change in ventilation air as it moves through a livestock or poultry building during hot weather.

change and moisture content change of the air as it moves through the house.

Line OA would represent an undesirable and unlikely situation in a livestock or poultry house in which all the heat was absorbed by the ventilation air in the sensible form. This would be strictly an air heating process and the dry-bulb temperature increase would be maximum for the total amount of heat absorbed.

In contrast, the vertical line OC represents a *psychrometric* change in the ventilation air in which all of the heat is *absorbed* in the latent form. Thus, no increase of dry-bulb temperature takes place. This is possible but would be a coincidence if the dry-bulb temperature change was exactly zero.

Line OB in figure elsewhere in this chapter represents the most typical situation between OA and OC in which some part of the total heat is absorbed as latent heat and some as sensible heat. The steeper and shorter the line the less *drybulb* temperature increase. Increased rate of air exchange through the building (cfm/bird or animal) will tend to shorten the line.

Line OD might represent any condition change from the vertical line OC to the left, which is most desirable. It means a decrease in dry-bulb temperature. This change of condition line OD can never be as far left as the wet-bulb lines on the *psychrometric* chart. This would represent an *adiabatic* saturating process. This evaporation process cannot technically be adiabatic, as for an *evaporative* cooler, because the ventilation air must absorb all of the heat produced in the building before any sensible heat of the air is converted to latent heat.

Figure elsewhere in this chapter shows the effect of *variation* of air flow in cfm/laying hen and the evaporative process taking place in the building, whether it be OA, OB, OC, or OD.

ARTIFICIAL COOLING OF LIVESTOCK

The possibilities of artificially increasing the comfort, or more importantly, the feed use efficiency of farm livestock is great. Adverse effects are apparent sooner from high than from low temperatures. In general, livestock are considered to be depressed by temperatures over 75 °F. There are, however, various ways of providing some relief for animals under heat stress conditions.

Drinking Water

Cooled drinking water proved of value in 6 yr of tests run on feeding beef cattle in California's Imperial Valley. Animals having access to cooled drinking water had an increased daily average gain of 0.30 lb.

There was found to be little difference between water temperatures of 60° and 70°F. Water consumption decreased slightly when the water was cooled. Cooling water only by evaporation did not lower its temperature enough to produce increased gains in beef cattle.

Air Movement

Fan movement of air in open feed lots for beef cattle in the Imperial Valley of California obtained some very startling results. In 1955 the average daily gain of the "fanned" animals was 44% more than that for the "*unfanned*" animals; in 1956 it was 22% more.

These gains were accomplished with 30⁰10 and 19% less feed per 100 lb of gain respectively. The difference between the 2 yr was due to better gains by the controls in 1956, possibly because of a generally cooler summer than 1955.

Large diameter (42-in.) slow speed fans were located at 50 to 75 ft spacing around the fence line. The fan center lines were 8 ft above ground and the fans were tilted downward at an angle of 7° toward the animals.

The rate of convective heat loss increases with approximately the 0.5 or 0.6 power of air velocity. The first increment of increased cooling by fans (*forced convection*), therefore, will be obtained with less cost than further increases.

Cooled Slab

Swine have been observed to lie down as much as 22 hr per day during hot weather. About 20% of the hog's surface area comes in contact with the floor when lying on its side. Tests were made at El Centro, Calif. in 1956 with the slab temperature at 65 °F and in 1957 with it at 70°F.

Control hogs were kept in an identical pen, but with a concrete unshaded water wallow instead of a cooled slab. There was shade near the wallow. In 1956 the pigs on the cold slab did not gain as rapidly as the control pigs. In 1957 the slab appeared to be beneficial. In 1956 the animals were a docile strain of Duroc-Landrace cross which were apparently content to remain on the slab thus spending little time at the feed trough.

The control pigs, due to discomfort, were forced to make regular trips to the wallow and were observed to spend more time eating. In 1957 the slab temperature was raised to 70°F in order to increase feed intake. The grade Yorkshire hogs used were considerably more restless and spent more time at the feed trough.

Possibly the most interest from a physiologist's standpoint was that the pigs on the slab appeared to be more comfortable during both years. Respiration rates ranged from 34 per min at 90°F air temperature to 68 per min at 105°F; whereas, the pigs with wallows averaged 140 and 158, respectively. Body temperature of the pigs on the slab averaged 103.4°F with 105 °F air temperature while the others averaged 105°F.

From an engineering standpoint the heat loss to the slab was most interesting. This is shown by figure elsewhere in this chapter. The rate of heat loss to the slab was increased slightly by the increase in slab surface temperature.

This indicates that thermal conduction of heat from the inner structures of the animal's body plays a minor part in heat loss, or a greater temperature difference potential would increase heat transfer. Heat transfer to the exterior surfaces of the animal is mainly due to blood circulation. Skin blood flow was evidently reduced by the cooler slab.

Inspired-Air Cooling

Marked responses to 2 levels of cooled inspired air (60° and 50°F) were obtained from dairy cattle subject to an ambient air temperature at 85°F and 50% RH at the Missouri Psychroenergetic Laboratory. Responses included increased feed intake and milk production, with accompanying decreases in rectal temperature and respiration rate.

Although the values of each parameter approached "*normal*" they failed to reach it, indicating that only a cooler total environment would provide maximum *relief*. Cooling capacity of 1/3 ton per cow was provided. For total barn air conditioning upwards of 2/3 ton per cow would be required for equivalent conditions.

Air Conditioning

Tests in the Imperial Valley of California indicated no economical advantage for air conditioning fattening hogs as compared to a shaded wallow. In one case feed and water were provided in the air conditioned house and the other outside. Hogs did occupy the air conditioned house more of the time than the *control hogs* occupied the shaded wallow, but this did not alter the final results appreciably.

The necessity of recirculating the cooled air to maintain 70°F in the cooled house presented a problem. The amount of dust collected on the return air filters was considerable. When the hogs were fed outside and weighed 130 lb, 20 gm/day of dust was collected on the filter, when they weighed 200 lb, 40 gm/day was collected.

When feed and water were provided inside, the dust collection was 41 gm and 70 gm/day at these 2 hog weights. These tests and others emphasize that not all air conditioning is beneficial, economical or trouble free. There are benefits from air conditioning to be derived, however only through careful design, operation, and management.

High temperatures rather than low temperatures present the major problems of environmental control and cause the greatest economical production losses. Environmental research in general shows the value of optimum temperatures and *humidities*.

Air conditioning will no doubt in the future become economical for many types of domestic animal enterprises; summer production of meat, milk, and eggs might be increased in many areas, and introduced in other areas where high summer temperatures have made production unsuitable.

No blanket recommendations for air conditioning can be made as to type of animal enterprise or climatic zone; each producer must weigh the proven benefits against the cost of equipment and operation.

In cooling design the total heat load of the building must be calculated in a similar way and possibly even more carefully than when designing the ventilation air exchange system for cold weather. This includes:

(1) heat transmission through exposed surfaces due to temperature difference and solar energy;

(2) heat and moisture released by hogs, cows, chickens, or other animals;

(3) ventilation air including infiltration air (sensible and latent heat); and

(4) heat from lights and equipment.

The *physiological* response time of heat-stressed swine to several cooling media can be summarized as follows:

(1) *Drinking cool* water by heat stressed pigs will bring about notable reductions in body and surface temperatures and in respiration and pulse rates within 5 to 15 min. The reduction of measured *stresses* was maximum within 15 to 20 min. Within 26 to 30 min all responses had started to increase again, and within 1 hr they were generally back to their beginning level.

(2) When water was applied to the *animal's surface*, reduction of body temperature was maximum in about 1 hr, and in 1.5 hr it was starting to rise again. It was about 3.5 hr before the body temperature had returned to its original value. Respiration and pulse rates had time responses similar to that for drinking water.

Evaporative Cooling

With properly designed pads and air flows below 200 fpm the dry-bulb temperature of the incoming air can be reduced to within 3°F of the wetbulb temperature. A more nearly complete saturation is attained with the lower air *velocities* through the pad. Figure elsewhere in this chapter gives results of tests with a 2-in. thick pad of aspen shavings.

The absolute values of evaporative cooling would vary some depending on pad thickness, *pressure drop*, and material. *Commercial fan literature* should indicate the size of pad required for specific fans. Field tests at Purdue showed a maximum dry-bulb cooling of 18°F on 1 particular day as indicated by figure elsewhere in this chapter.

6

ENVIRONMENTAL CONTROL AND THE CLIMATE

Cold weather air exchange systems for livestock and *poultry structures* must remove excess *vaporized moisture*. If enough air is moved through the building to remove the respiratory and body surface vaporization, normally other requirements of ventilation air exchange (*oxygen supply*, removal of carbon dioxide, *ammonia*, *dust*, and *odors*) will be adequately satisfied.

Removal of vaporized moisture will prevent cumulation and condensation in the building and maintenance of an optimum relative humidity below 80%. Vaporized moisture production data for domestic animals and birds are presented in other chapter of this book.

From figure elsewhere in this chapter the air exchange required to remove 1 lb of water vapor per hour may be determined. Thus, if the pounds of water vapor production per hour are known for the house, the minimum ventilation requirement may be calculated.

DESIGN CONDITIONS

The cold weather outside design temperature must be selected according to the climatic conditions of the locality in question. The average temperature for the *coldest month* of the year is a *justifiable* and easily determined basis for the outside cold-weather design temperature. Figure elsewhere in this chapter gives the average January temperatures for the United States.

For example, in central 'Michigan the average January temperature is approximately 20°F. Thus, during this coldest month there must be enough air exchange to prevent any net accumulation of vaporized moisture in the livestock or poultry building. One-half of the time when it is colder than the mean temperature there will be temporary accumulation, and one-half of the time additional drying.

This fluctuation may be no longer than for a diurnal cycle. It might often be below the mean a few hours at night and about a few hours during the day. Considerable experience with poultry housing in the northern United States confirms the practicality of designing cold weather air exchange for the removal of vaporized moisture at this average-outside design temperature.

Example 6.1—A typical White Leghorn laying hen will produce about ¼ lb of vaporized moisture per day at inside housing temperatures of 50°F. This is a per hour rate of slightly over 0.01 lb per bird. From figure elsewhere in this chapter at a 20°F outside design temperature and a 50°F inside design temperature, a minimum of = ½ cfm (0.01 × 50) will be required to keep at or below 80% RH. The thermal heat balance of the building must then be checked to determine its adequacy for providing this minimum air exchange of ½ cfm per bird under these conditions.

Example 6.2—A finishing house for 100-lb hogs may be checked for the outside design temperature in a similar way. A 100-lb hog will produce 0.155 lb of vaporized moisture per hr. Thus at 50°F inside temperature and 20°F outside 7.75 cfm (0.155 x 50) air exchange per hog is required to keep the inside at or below 80% RH.

Minimum Air Exchange

Ideally there should be some continuous exchange of air through a livestock or poultry building at all times. About 50% of the time during the coldest month the design air exchange for vaporized moisture rem©oval will not be possible.

The air exchange control system should provide an absolute minimum of the design air exchange for all temperatures that prevail 97.5% or more of the time. Figure elsewhere in this chapter shows the winter temperatures that prevail for the United States less than 2.5% of the time.

For example 6.1, the poultry house should be checked for 1°F, which is the low temperature exceeded only 2.5% of the time in Detroit, Michigan. At this temperature there should be available sensible heat to provide ¼ cfm; bird continuously. (This is ¼ of the minimum water vapor removal air exchange rate of 1 cfm/bird.)

Example 6.3—For a 10,000 bird laying house there should be a minimum fan capacity of 2,500 cfm (¼ × 10,000) properly controlled thermostatically to provide this minimum air exchange continuously all but 2.5% of the cold weather period. During the short periods that make up the 2.5% of the cold season, the themostatically controlled 2,500 cfm fan would periodically turn on automatically for short periods to provide minimum fresh air requirements.

It is estimated that it would take approximately 9 hr of zero air exchange in a typical 10,000 bird laying house to deplete the oxygen supply to a dangerous level of between 16 and 6%. Below 16% is undesirable, and 6% is the killing level. This is based upon the respiration air requirement of hens as follows

$$\text{cfm} = (0.0001185) \times (\text{body wt}) \times (\text{respiration rate per min}) \quad (6.1)$$

A 4-lb White Leghorn breathing 40 times per minute would need 0.019 cfm for normal respiration. The air breathed is depleted 5% in oxygen. From the poultry house volume available per hen, the maximum zero-air-exchange period can be estimated.

This assumes absolute air tightness which even windowless structures generally do not have. In any case, if the birds or animals can survive the cold inactive night hours, the excess heat generated in the morning by activity and feeding will always cause the thermostats to activate the minimum-air-exchange fans.

Time Clock Control

An overriding time clock on the thermostatically controlled fans can provide the assurance of minimum airflow regardless of outside or inside temperatures. With a properly designed and operated house and ventilation system for adult birds or animals, the time clock has not proved justifiable.

With a time clock, control of the inside minimum *temperature* is lost. It also provides a *tendency* for some operators to attempt to prejudge weather conditions daily and then set the operating time period on the time clock accordingly. This leads to trouble, besides nullifying all advantages of the automatic thermostatic equipment.

Winter Sunshine

Livestock and poultry structures designed to absorb solar radiation during cold *weather months* as a source of supplemental heat have been studied. The availability of solar radiation in a specific area where a solar designed house is being considered should be given primary attention.

Figure elsewhere in this chapter shows a map of the United States which indicates average hours of sunshine daily for the winter months (December, January and February).

It is noted, in particular, that the northwestern part of the United States and the Great Lakes region have an average of only 3 hr of sunshine daily. Solar design houses would be questionable in these areas.

Based upon daily hours of sunshine available, radiation intensity, and predicted increased heat to the building for improving environment, the *economics* of investment versus returns must be determined. Additional structural costs must be offset by additional production returns from the enterprise.

Exposure Factor Determination

The exposure factor [E.F. = $U_{av}A/N$ (equation 6.1)] was discussed in the previous chapter for use in determining possible air flow when a known quantity of heat is available and a specific building is being considered. The *adequacy* of the possible air exchange may then be checked for various inside and outside temperature *combinations*.

The *exposure* factor may also be used more directly as a design criterion. The first question in design of a new building is what exposure factor should it have. In other words how much insulation should be provided for in the *sidewalls* and *ceiling*.

Once the inside and outside design temperatures and a necessary minimum air exchange have been established the exposure factor can be determined. For this purpose the exposure factor is

$$\text{E.F.} = \frac{q_s - (0.16)(v)(\Delta t)}{\Delta t} \qquad (6.1)$$

where

q_s = Sensible heat production per bird per hr

0.016 = specific heat of air in Btu/(ft^2)(°F)

v = ventilation air exchange in cu ft/(bird) hour)

Δt = temperature difference between inside and outside

It will be noted that development of an exposure factor using this formula (equation 6.1) will provide only for removal of respired moisture at the outside design temperature. This should be the average temperature in that area for the coldest winter month.

Example 6.4—If the results of the example 6.2 problem are used, a necessary exposure factor may be determined. In this example for t_i = 50°F and t = 20°F it was found that 7.75 cfm of ventilation air exchange was necessary to remove the 0.155 lb of water directly vaporized per hour by the 10000-lb. hog. From figure elsewhere in this chapter the total sensible heat is 510 Butcher and from figure elsewhere in this chapter the latent heat is 160 Btu/hr thus

$$E.F. = \frac{Q_s - (0.016)(v)(\Delta t)}{\Delta t}$$

$$E.F = \frac{350 - (0.016)(7.75)(30)(60)}{30}$$

$$E.F. = \frac{350 - 223}{30}$$

$$E.F. = 4.23 \text{ Btu/(hr) (°F) (hog)}$$

Now having a necessary exposure factor, the amount of insulation (U_{av}) required may be determined from the original definition of the exposure factor (E.F. = $U_{av}A/N$) once a building size and number of animals to be housed have been established.

If the resulting insulation requirements are beyond reason then more animals and/or supplemental heat will be necessary for optimum conditions.

HOT WEATHER AIR EXCHANGE

Air exchange must be sufficient to keep the inside temperature at or close to outside temperature during hot weather. *Summer climatic* conditions will dictate the need for amount and control of air exchange. The first consideration is normally the prevailing dry-bulb temperature for a specific locality. Figure elsewhere in this chapter gives the temperatures throughout the United States which are exceeded only about 75 hr for the summer months.

The temperature-humidity index THI (or discomfort index DI) provides a more valuable measure of hot weather climates. This index includes the effect of relative humidity along with dry-bulb temperature. Figure elsewhere in this chapter shows the average discomfort index at noon for July in the United States.

Figure elsewhere in this chapter shows a somewhat lower average July discomfort index basedon observations at 1:30 A.M.. 5:30 A.M., 1:30 P.m., and 7:30 P.M. EST. These DI values would be more nearly daily means than for figure elsewhere in this chapter which represents one noontime measurement.

Milk Production in Hot Weather

Of primary importance to the producer, and thus to an engineer considering environmental design of a dairy shelter, is the effect of temperature and humidity on milk production. Figure elsewhere in this chapter presents the results of a study which relates milk production decline to the DI for *Holstein cows.*

It will be Noted from the production decline curves that the heavy (50 lb/day) producers will decline more in pounds of milk per day at higher DI than the 30-or 40-lb cows. Actually, the higher producing cows still give a greater amount of milk at DI of 80, though undoubtedly at a sacrifice to future production.

The 50 lb per day cows decline about 1.7 lb per day for each unit increase of DI, whereas the 30 lb per day cows decline about 0.7 per day for each unit increase of DI above 74.5.

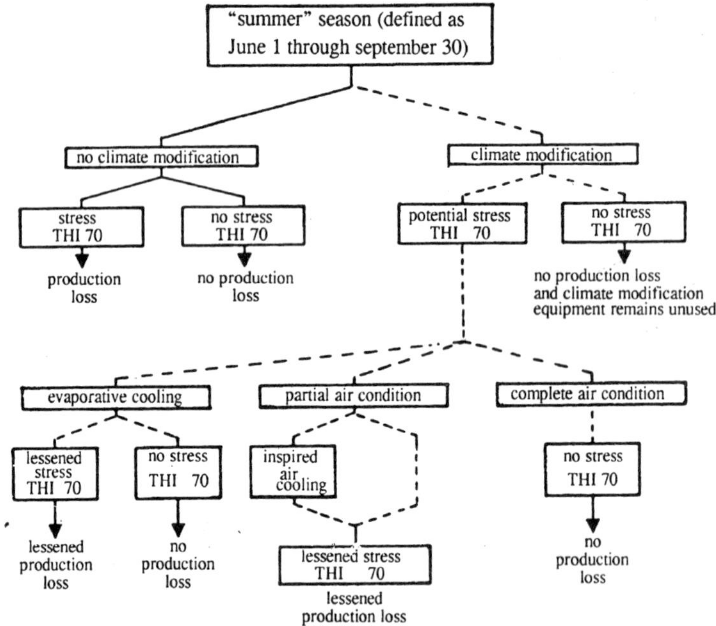

Figure 6.1: Sequence of decisions and consequent events to illustrate alternatives of climate modification for dairy heard managment.

The sequence of decisions for the case of climatic modification based upon DI and the resulting events are outlined in figure elsewhere in this chapter. One decision is to make no attempt to modify climate. From figure elsewhere in this chapter the *probable production* losses for dairy cows can be predicted.

If the producer decides to provide shelters and related equipment, he still must decide on the type of modification. Any modification must be justified by increased production.

Climate modification can range from minimum shades to completely enclosed shelters with total air conditioning.

The probability of many different weather events can be estimated from climatological data, such that

$$P = P[f(W)]$$

where P = the probability that event W lies between two chosen values and is a function of the occurrence of that event

In this case the critical weather event of concern is DI; however, it might be temperature only or other parameters related to the *biological* response. *Combining* the obtained probability of a certain weather event with the biological response function related to that event provides the probability of the given biological response

$$P(B) = P[f(W)] \tag{6.3}$$

From the probabilities of penalties in the form of decreased milk production, the manager can decide on the degree of climatic modification. Estimates of the probability of *exceeding* various DI levels for a 20-yr period at Columbia, Missouri, were examined. In figure elsewhere in this chapter the empirical *probability* DI curve as observed on an hourly basis is related to the biological response (production decline) curves of figure elsewhere in this chapter in the *monogram* on the right-hand side. This allows direct reading of

$$P(B) = P[(f(W)]$$

Only daily mean DI values greater than 70 have been shown to be detrimental to production, therefore the region above DI = 70 is of primary interest.

The horizontal dashed line shows the *probability* of a 1.0 lb per day loss for a 50 lb per day cow to be about 0.32. This 1.0 lb loss would occur about 1 day in 3 during the summer months. The vertical dashed line shows that the corresponding daily mean DI value of 73.5 will cause a production *decline* of 1 lb per day.

The DI probability curve further shows that about 40% of the summer days in *Columbia* will cause no production loss for 40- and 50-lb producers. For the June 1 to Sept. 30 season it may be further calculated that the following production losses should be expected

51.0 lb for 30 lb/day production level

87.6 lb for 40 lb/day production level

127.0 lb for 50 lb/day production level

These examples illustrate how management decisions can be made from empirical probabilities.

Economic Analysis

An economic analysis was made of summer environmental control for livestock based on the laboratory findings of milk reduction. The report presented the results of an investigation of costs and returns of summer air conditioning of dairy barns, based on 16 locations in the United States. Records were studied for 122-day periods from June 1 through Sept. 30.

To establish the costs and returns associated with environmental control, functional relationships are required between weather-dependent factors and a suitable climatic parameter. For Holstein dairy cows, milk production decline has been related to the daily mean temperature humidity index (THI) value through the equation

$$MDEC = -2.370 - 1.736\ NL + 0.02474\ (NL)\ (THI) \qquad (6.2)$$

where MDEC = absolute decline in milk production, Ib (day) (cow)

NL = normal level of production, lb/day

THI = daily mean temperature humidity index

Under environmental control, these declines will not occur and are viewed as gains in the economic analysis. Similarly, feed intake decline has been related to the daily mean THI value through equation

$$HDEC = -62.24 + 0.863\ THI \qquad (6.3)$$

where HDEC = absolute decline in hay consumption, lb/(day) (cow)

THI = daily mean temperature—humidity index

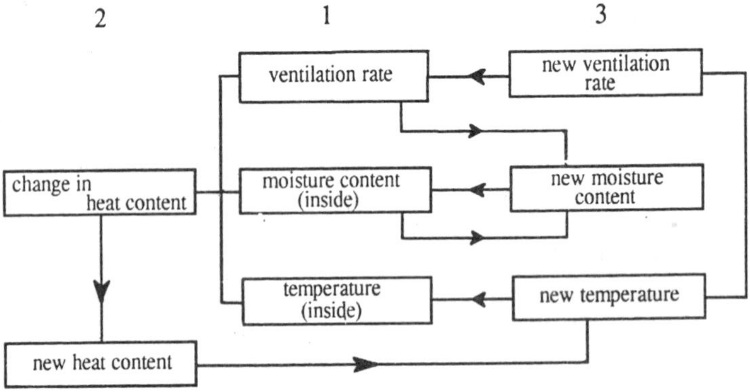

Figure 6.2 : Flow daigram for heat and moisture change in the ventilation system model.

Because of environmental control, additional feed is consumed and represents a cost factor.

The cost factor of *electric energy* used by air-conditioned farm structures has been related to cooling degree days through the equation

$$EC = 2.8{:}3 + 0.68DD \quad (14.4)$$

where EC = daily energy consumption, kwh/ton of air conditioning (assumes 1 ton refrigeration requires 1 hp input)

DD = daily cooling—days, computed using a mean daily dry-bulb temperature base of 60°F

From the climatological records for a given location and period of time, *empirical probabilities* for weather events were computed. This allowed calculations of the expected number of days per season associated with each climatic variable. When combined with the associated production, *feed intake*, or energy use value and integrated over the season, seasonal values for the variable were obtained.

To judge the potential of air-conditioned dairy barns in the United States, input values were *arbitrarily* selected so that an isoline of *profitability* was determined for cows with a 70 lb per day production level.

The input values and the isoline are shown by figure elsewhere in this chapter. Below this line indicates an area where environmental control should be *investigated*, particularly for high producing cows.

Evaporative Cooling

This method of cooling is most efficient where there are low wet-bulb and high dry-bulb temperatures. Figure elsewhere in this chapter shows the *distribution* of wet-bulb temperatures throughout the United States that will be exceeded not more than 5% of the total hours during June to *September inclusive*.

Southern Arizona has 100°F dry-bulb and 70°F wet-bulb, where central Michigan has 85°F dry-bulb and 70°F wet-bulb temperatures. An efficient pad and fan *evaporative cooling* system can decrease the dry-bulb temperature of the incoming air adiabatically to within 2 or 3 ° of the wetbulb temperature.

The amount of dry-bulb cooling can thus be estimated. In Arizona the decrease in temperature would be over 20°F while in Michigan it would be only slightly more than 10°F. The value of the temperature reductions in production increases must then be evaluated.

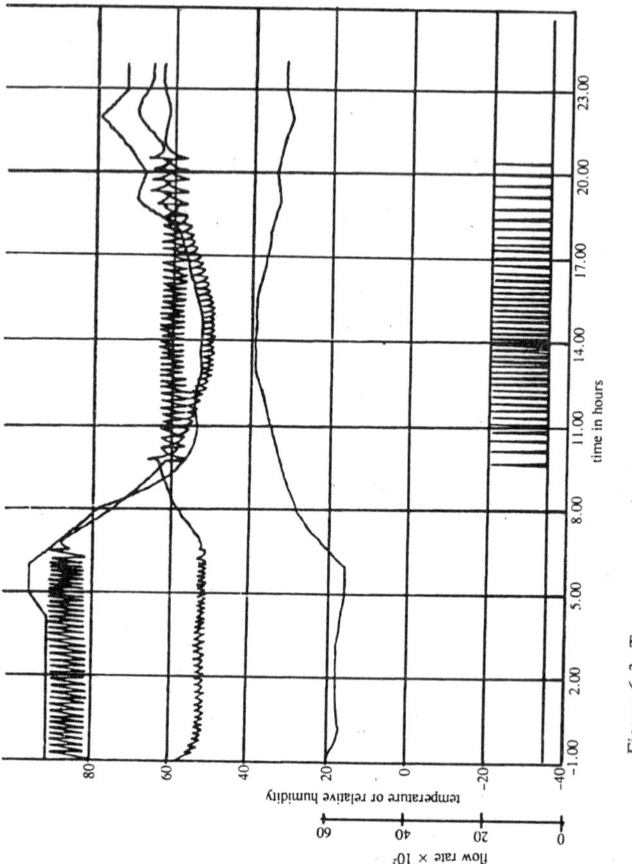

Figure 6.3: Temperature and relative humidity record for thethermostat control system with two fans.

MODELING OF VENTILATION CONTROL SYSTEM

Scientists devised a mathematical model of the environmental control variables and adapted it to digital computer language. The basic equations were obtained by heat and *moisture balances*. The flow chart of figure elsewhere in this chapter shows the *mechanics* of the program. The following conditions were assumed.

Given and constant: time interval, heat produced, moisture produced, mass of air in building

Given initial conditions: inside temperature and moisture content, ventilation rate, heat content

.eau in at each time increment: outside dry-bulb and dew-point temperature

Figure elsewhere in this chapter shows the plotted results from the Calcomp computer plotter. One particular control system of two *thermostatically* controlled fans was assumed for this example swine structure. The *minimum* flow rate with one small fan was not adequate during nighttime temperature of less than $t_o = 20°F$ to keep the inside relative humidity RH, below 80%.

Any greater amount of air flow would, however, tend to lower the inside temperature t_i, possibly below the desirable level. When the outside temperature t_o increased to above 30°F during the forenoon, the heat balance allowed the second fan to turn on *intermittently*. Satisfactory conditions inside were then maintained.

7

CATTLE

It is estimated by F.A.O. (1962) that in 1960-61 there were approximately 930 million cattle in the world. If the term 'tropics' is used in its broadest sense then something more than half of all these cattle were raised in that area, but it is likely that this is an overestimate of the number raised in the geographical tropical area as the estimates for Africa, Asia and Australia include some cattle raised in the temperate or sub-tropical area. Some 40 per cent of the cattle raised in the tropics are in India and 25 per cent in Africa.

Their place in the life of the indigenous peoples of the tropics often differs remarkably from the position they are given by man in the temperate zones. In the tropics, as elsewhere, cattle are essential as a source of food, work and of many by-products of great value; but with millions of people these considerations, though vital, are secondary to the part cattle play in religion, in social custom, as a reserve of family wealth and as a mark of respectability.

The nomad would as soon deplete his herd as the family man in Europe would his capital account; the Hindu would rather starve to death than eat his cow; and in different ways and to different degrees there are few tropical people who have not some regard for cattle quite apart from their immediate economic use.

Tothill thus describes the outlook of the Dinka of the Sudan: 'As a result of living in such close contact with their cattle, their animals have come to assume an almost religious significance with them. Every self-respecting young man possesses a special bullock in which his personality is alleged to be symbolized; these animals are specially fed and groomed, their horns are trained and decorated with tassels, they are lauded in song and regarded with an affection almost amounting to worship.'

Table 7.1 : Estimated cattle population, human population and human population per head of cattle in temperate and trópical climatic areas.

Area	*Number of cattle 1945-61*	*population 1961*	*Human population per head of*
Temperate	Thousands	Thousands	
North America	150,431	202,116	1.3
South America	59,585	28,192	0.6
Europe	117,100	428,000	3.7
U.S.S.R.	75,780	218,000	2.9
Asia (including China)	54,561	779,429	14.5
New Zealand	6,446	2,422	0.4
Total	463,903	1,658,159	
Tropical			
Central America and Caribbean	35,301	69,008	1.9
South America	106,540	11,765	0.1
Africa	114,100	261,300	2.4
Asia	193,629	689,783	3.1
Oceania	17,654	13,778	0.8
Total	467,224	1,045,634	
World Total	**931,127**	**2,703,793**	

ORIGIN

The group of animals that includes all types of domestic cattle are known as the *Bovidoe*. Today it is usual to list four sub-groups, each including several species and many breeds.

(a) The Taurine sub-group comprises the majority of domestic cattle including the largest and most important species, *Bos taurus,* the European type, and *B. indicus* or the zebu.

(b) The Bibovine sub-group, which includes the gaur *(B. gaurus), a* wild species of India, Burma and Malaya; the gayal *(B. frontalis),* a semi-domesticated species found in Assam and Burma; and the banteng *(B. sondaicus),* which is an important source of beef in Indonesia, Borneo and Malaya.

(c) The Bisontine sub-group that includes the yak *(B. grunniens)*, an important domestic animal in Tibet, and the North American bison *(B. bison)*, a species that had almost become extinct but is now increasing in numbers and has been crossed with European cattle to produce the crossbred known as the 'cattelo'.

(d) The Bubaline sub-group or buffaloes. The important species are *B. bubalis* found in the Near East and South-east Asia. *B. mindorensis* found in the Philippines, and the Celebes buffalo *(B. depressicornis)*. These are important domestic animals in the tropics and are discussed in a separate section.

The great majority, but not all, of the indigenous tropical breeds of domestic oxen are zebus of one variety or another, that is, they are humped. It appears that the earliest ancestors of modern cattle were those domesticated in central Asia about 8,000 B.C., and that a now extinct variety *of Bovidae,* which at about the same period in history is known to have existed in India, also may have contributed.

There is evidence that the cattle domesticated in South Turkmenistan were long-horned, and that a variety of smaller, short-horned cattle of the *B. nomadicus* type was derived from them. Domestic cattle had reached Babylon by 5,000 B.C., and Egypt before 4,000 B.C.

There is evidence that in India, certainly from 3,000 B.C., there were two types of cattle: a large, long-horned, humped variety and a smaller, short-horned type. There is nothing to indicate how the wild, humpless animal developed into the domesticated humped one; all we know is that the zebu carries traits of three varieties of the genus *Bos,* that it probably developed in the Indus valley, was specialized in the Indian sub-continent and spread through southern Asia.

Olver describes the characters of what he considered to be five basic types of the Indian zebu, the distribution of each being associated with human migration into and across the continent. The earliest appears to be the small, black, red and dun cattle found in many hilly tracts and forest areas throughout India and which resemble the original *B. brachyceros,* except that they are humped.

The Gir type, which are domiciled in western India, are characterized by a prominent, convex broad forehead, but have the long head of the typical zebu with comparatively thick horns emerging from the side of the poll and running down and backwards. The Mysore type found in the south-west, have a long narrow face, a forehead bulging above each eye, and horns which emerge close together from the top of the poll and run up and backwards. The large grey-white type of cattle have what are now considered to be the typical zebu characteristics.

There are two varieties, the large, broad-faced, lyre-homed oxen, reared along the Aravalli hill range to the west, and of which the Kankrej breed is the outstanding example, and the narrow-faced, short-homed type which are found through northern and central India and as far south as Madras, of which the Hariana and Ongole are examples. The fifth type is expressed by a breed in the north-east Punjab known as the Dhanni.

In Southeast Asia many of the cattle are humped, but they are not entirely of the zebu type. It is likely that most of them are the result of a mixture of zebu and Shorthorn *(Bos brachyceros)* types. Generally those of the countries closest to the Indian Peninsula exhibit the most zebu genes and those in the countries most distant, such as the Philippines, being almost entirely of the shorthorn type.

In addition to the domestic cattle there are numbers of wild or semi-wild cattle of other genera, the most important of these being the gaur, the gayal and the banteng. A new genus, known as the kouprey *(Novibos* spp.), has been discovered in the forests of Cambodia.

Whether cattle were domesticated in prehistoric North Africa is a controversial question, but it is known that three major types of cattle were imported or migrated with nomadic people into northeast Africa and that these are the ancestors of the vast majority of cattle found in Africa today.

The history of the domestication of cattle in Africa has been clarified by Payne, who states that the immigrant cattle were the Hamitic Longhorn and the Shorthorn, considered ancestral *Bos taurus* types, and the zebu of the *Bos indicus* type. These mixed at different times and in different ways to create Sanga cattle-the dominant type in Africa today.

Although Sanga cattle resulted from crossbreeding at many centres in East, North, West and Central Africa, East Africa was probably the main centre of origin of all the Sanga breeds now found in East, Central and South Africa.

Hamitic Longhorns were the first arrivals and Faulkner and Epstein state that the earliest records of their presence in Egypt are dated about 5000 B.C., though this date should be accepted with caution. They were followed about 2750-2500 B.C. by the humpless Shorthorn cattle *(Bos brachyceros)* and the peoples owning the Hamitic Longhorn were pushed or migrated voluntarily westwards and southwards, so that today there are no traces of the Hamitic longhorn in Egypt.

There were possibly three main migration routes: westwards along the Mediterranean littoral, south-westwards via Tibesti and Tassili highland areas, and southwards up the Nile towards the highland areas of what are today Ethiopia and Kenya.

When the westward migration of Hamitic Longhorn cattleowning people reached what is now Morocco, the stream probably divided, one branch proceeding north into Spain, Portugal and beyond; the other west and southwards into West Africa.

Curson and Thornton have suggested that descendants of the northern migration of cattle can be seen today in such breeds as the Raza de Barroza of Portugal and the Andalusian cattle of Spain. Cattle of this type were exported from the Iberian Peninsula and are represended today in America by such breeds as the Franqueiro of Brazil and the Longhorn of Mexico.

A breed descended from the Hamitic Longhorn still exists as a more or less pure type along the Gulf of Guinea and is known as the N'dama. It has apparently been preserved because it developed or exhibited a high degree of resistance to trypanosomiasis; in consequence, it has been able to penetrate the forest areas.

A second Hamitic Longhorn found in West Africa is the Buduma or Kuri. It probably owes its survival as a separate breed to the fact that it has been isolated on the islands of Lake Chad.

The southward migration of Hamitic Longhorn cattle is not so easy to trace, but early rock, paintings and engravings in Ethiopia, Somalia and as far south as Mount Elgon in Kenya suggest that the Hamitic Longhorn was the first domesticated animal to enter this region, and the remnant breed known as the Binga in north-west Rhodesia appears to be a dwarf form of Hamitic Longhorn which apparently penetrated far to the south.

The Shorthorn-type cattle followed the Hamitic Longhorn fanning out along the northern Mediterranean littoral and southwards along the Nile into Upper Egypt and the Sudan, but not across what is now the Sahara, presumably due to increasing aridity of the region.

Modern cattle in Egypt exhibit characteristics that have been inherited mainly from these Shorthorn immigrants, while existing breeds in Libya, Tunis, Algeria and Morocco are almost entirely of the Shorthorn type. They probably split into two migratory streams in Morocco, one moved north into the Iberian Peninsula and into France and the British

Isles; the Jersey, Guernsey and Kerry breeds apparently being partly derived from this stock. The other stream moved southward and westward until it came to the forest barrier. Like the Hamitic Longhorn preceding them, these animals gradually acquired a resistance to trypanosomiasis, and are known today as the West African Dwarf Shorthorn.

Their resistance to trypanosomiasis is not as marked as that of the N'dama. Little is known of the southward movement of Shorthorn cattle, though according to Faulkner and Epstein a small number of cattle of this type are still to be found in the Koalib hills in the Nuba mountains of Sudan and there is evidence that they existed in Uganda within historical times.

There is also every likelihood that Shorthorn-type cattle were imported independently into the Horn of Africa, the East Coast of Africa and the offshore islands.

The chest-humped zebu is a comparative newcomer to Africa. It is likely that cattle of this type were imported in small numbers into Egypt as early as 2000-1500 B.C. and into the Horn of Africa and possibly East Africa, but it is unlikely that they arrived in large numbers until after the Arab invasion (c. A.D. 669).

Faulkner and Epstein consider that the neck-humped zebu was introduced into Egypt during the period 1580-1265 B.C., but as Slijper has shown that crosses between humpless cattle and chest-humped zebu cattle tend to possess neck humps, it is likely that these cattle were some of the first Sanga type.

Zebu cattle appear to be more resistant to rinderpest than humpless or Sanga cattle, and are also better adapted to living in arid and semi-arid areas. They have therefore spread rapidly in these regions and where rinderpest has been epidemic and today they are found in West Africa, where they are represented by the White Fulani; and as far south as Malawi and Zambia.

This dynamic expansion has not yet spent itself and the zebu is still gradually replacing the Sanga and other breeds along the border where they meet in Uganda, northeast Congo and northern Ethiopia.

The term 'Sanga' has been used in the past either to describe a particular type of cattle such as the West African Sanga or in the broadest sense to describe any breed that is apparently the result of a mixture of Zebu and Hamitic Longhorn and/or Shorthorn.

Mason and Maule suggested that the so-called zebu breeds with cervico-thoracic or neck humps should be classified as Sanga. This suggestion should be carried to its logical conclusion and all crossbreds between humped and humpless cattle in Africa should be termed Sanga. If this definition is accepted, the possibilities for the formation of different types of breeds of Sanga in Africa, over the period of recorded history, are very great, with tribal migration, raiding and counter-raiding maintaining the gene mixture in a state of flux.

The available evidence suggests that it is likely that East Africa and Ethiopia were centres of origin of large numbers of the earliest Sanga cattle and that these would be neck-humped animals. Migration may have taken place northwards, westwards and southwards. There had been pressures on the early inhabitants of this region to migrate with their cattle southwards, thus the neck-humped Sanga would be found farthest to the south, as indeed they are.

The choice of migration routes southward was controlled by the availability of feed and water and the location of the tsetse *(Glossina* spp.) country. One or more fly-free corridors undoubtedly existed between East and South Africa.

The first migration southwards by Hottentot-type people, about A.D. 1400, probably followed a route between Lakes Tanganyika and Nyasa to the Upper Limpopo, then westward crossing the Orange River, and ultimately meeting the first Europeans advancing north from the Cape. The Sanga cattle owned by these early migrants are the ancestors of what are known today as Africander cattle.

Migrating behind the Hottentot people came the Bantu tribes, also herding Sanga cattle of various types. The majority of Sanga breeds in Zambia, Angola, Bechuanaland and South West Africa are of the Hamatic Longhorn x zebu type, while those of the central region of Zambia and the eastern region of Rhodesia were probably derived from the Shorthorn x zebu crosses.

In East Africa the Sanga cattle have been displaced by zebu cattle only in very recent times, and pockets of Sanga cattle are still to be found in the zebu areas.

Western migration of peoples herding Sanga cattle took place from Ethiopia to what is now Uganda and further west along the northern borders of the rain forest to West Africa, so that today it is likely that the Sanga cattle of West Africa may be derived from two sources; local crossing of immigrant zebu cattle with local Hamitic Longhorn and Shorthorn cattle, and immigrant Sanga stock from East Africa.

No cattle existed in the Americas or Australia when these continents were discovered by Europeans. The Spanish and the Portuguese introduced their own breeds to the American littorals. These spread over the islands in the Caribbean and inland throughout the tropical and sub-tropical zone on the mainland. In the nineteenth century British and other European breeds were introduced, and more recently small numbers of tropical type cattle have been imported from Africa and Asia.

Although much of the original European stock degenerated, at several centres in this area new breeds are evolving. Some, such as the Criollo and Caracfi from Spanish stock, the Achiote (Criollo × Shorthorn) in Guatemala, the Jamaica

Hope (Jersey × Sahiwal) in Jamaica, the Romo-Sinuana (Red Poll and/or Aberdeen Angus × Horned Sinu) in Colombia, and the Nelthrop (Senegal N'Dama × Red Poll) in Antigua from crossbreeding, and some, such as the Indo-Brazilian (Gir, Kankrej and Ongole) in Brazil from a mixture of *B. indicus* types.

In addition there are three new breeds in the sub-tropical area of North America that are of increasing importance; they are the American Brahman, the Santa Gertrudis and the Beefmaster and there are several other types of unstabilized and useful crossbreds.

In Australia, the first importations were made less than two hundred years ago. They were mostly African cattle but a few were from India. They were apparently neither well selected nor located and were soon replaced by British breeds, mostly Shorthorns.

As the importation was into the temperate zone, it was successful and the acclimatized British breeds were carried further and further north into the tropical zone with the extension of land utilization. In the north the British breeds were not so successful and recently American Brahman and Santa Gertrudis stock have been introduced from America and Red Sindhi stock from Pakistan. The tick resistance of zebu crossbreds has allowed great advances in coastal Queensland.

There are no polled breeds indigenous to tropical Asia and only one in Africa, that is, the zebu of Mbulu in Tanganyika (Epstein, 1955). The polled character is considered by many to be derived from the extinct *Leptobos*. Polled individuals appear from time to time and in considerable numbers in many of the Sanga type cattle, and loose-horned individuals are frequent both in them and in the zebu of certain types.

TYPES OF TROPICAL CATTLE

The average cattle owner in the tropics does not have sufficient resources to maintain types of cattle with functional differences and he requires an all-purpose beast which will work, reproduce and supply him with milk while alive and provide meat when killed. According to his crops, the size of his holding and his social status, the priority of his requirements varies to quite a considerable extent. It is not correct to say, as some do, that no functional specialization of tropical cattle has taken place, either recently or throughout the centuries, which can in any way be compared to that which the European breeds have been subjected.

It certainly has not been done with the same intensity of purpose nor has production in any direction been brought to the high level that it has reached in the temperate zones, and it is doubtful if it ever can be, but there is as great a difference between one tropical breed and another in their ability to produce a given commodity as there is between two differently constituted European breeds.

The Indian Sahiwal and Red Sindi in their own way, are as specialized dairy breeds as are Jerseys or Friesians; the Gir in India and the Boran in East Africa are outstandingly suitable for beef production in the tropics; the Indian Dhanni and Nagore are specially bred for fast work and the Kankrej and Krishna Valley for heavy draught; the Amrit Mahal of Mysore and the Rahaji of Nigeria were, in their day, and in different ways, specialized for war and pageantry.

In many tropical regions, however, the scarcity of water, the innutritious nature and the scant growth of vegetation available for fodder, the prevalence of highly, fatal disease and, up to last century, the danger of human marauders, left little or no opportunity or inducement for improvement in any direction and many millions of tropical cattle justify their existence merely by the ability to survive.

The original purpose of domestication was, doubtless, to facilitate or ensure the supply of meat, and that remains, probably, the main function of tropical cattle today, in spite of the fact that so large a part of the human population is composed of Hindus and other vegetarians; it is certainly the case in Africa, Arabia, tropical America and Australia. In South-east Asia, work is the primary function; in certain East African territories and in the richer agricultural regions of the Americas, it is one of the main functions, although machinery is quickly displacing draught animals. One cannot, however, dissociate work cattle from meat, for that is their ultimate end, as it is for all cattle.

CHARACTERS AND CONFORMATION

Beef Type

Tropical beef cattle are of two kinds; one which can convert large quantities of easily obtained fodder into fair quality meat, and the other which can gather and make use of sparse-growing, coarse fodder, turning it into meat which may be of low grade but which nevertheless is of considerable dietary value to man. Each is specialized to make the best of the food available and, until agricultural methods have been radically altered, each will be required in its own sphere.

The former have the characteristics of the specialized beef breeds in that the meat-carrying parts of the frame are well developed, while the extremities are just large enough to support it; the latter have the strong, long limbs required to cover great distances with ease. In India the two kinds are typified by the Gir and the Kankrej.

In East and South Africa by the Boran and Africander, and in West Africa by the Sokoto Gudali and the Red Bororo or Rahaji. In tropical Asia and Africa, where the people are poor and the price that can be paid for meat is low, where the home markets are not well organized and where export facilities do not exist, the present unimproved types of indigenous cattle satisfy requirements.

Few if any concentrates are available for beef production, and forage in one form or another is the sole food. Where that is plentiful and easily come by, the orthodox beef types are used; where it is scarce or its collection entails much work, the leggy types must be employed.

Under these conditions where price limitations severely restrict the means which can be used to produce beef, the advantages of early maturity, rapid growth and a quick turn-over have not the same importance as they have for a richer market.

The primary essentials for the current, rather haphazard husbandry is the ability to make the most of the very poor food available during the scarcity season of the year, and to reproduce regularly at the shortest intervals possible.

Where cattle are raised, as they are in America and in tropical Australia, to supply a keen export or an active home market where quality can be paid for and the demand is constant and unlimited, only stock of the best improved beef breeds is good enough.

In Africa the market for high-quality meat is supplied either from favourably situated ranches upon which the best indigenous breeds are reared or, in still more favourable circumstances, the European x indigenous crossbred is used.

European breeds have generally failed to withstand the climate in the American tropics, and pure zebus have often not reached the standard of productivity required but a combination of the two, as typified in the Santa Gertrudis, has been found very satisfactory in the Gulf Coast

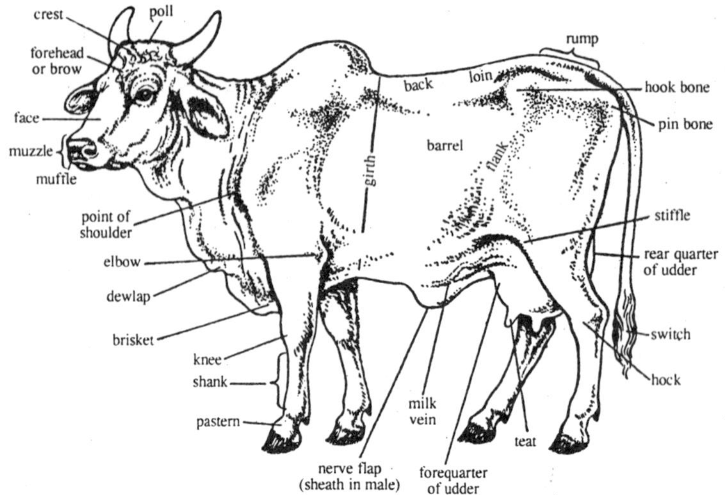

Figure 7.1: The salient points of cattle.

region of America. The Australian meat export trade still relies upon European breeds, even in Queensland and the Northern Territories but, as has been mentioned, in the most difficult regions, crossbreeding with the zebu and the Santa Gertrudis is in operation.

In countries where large-scale ranching is practised, the husbandry is so arranged that the losses which otherwise arise during the scarcity periods of the year are minimized by supplementary feeding on more favoured grazing areas, and all marketable stock are disposed of while they are still at their best.

The breeds from the temperate zone which have given the best results are the Shorthorn; Hereford and Devon. The zebus most favoured have been Ongole, Gir and Kankrej, and the *American Brahman*.

Work Type

The characters required of good work cattle are not only great strength, free-moving limbs and sound feet but a quiet temperament and submission to discipline. There are few breeds of zebu that cannot be used for work, but many are defective in one respect or another.

Their use in heavy road haulage is yearly becoming more restricted and the large types such as the Indian Kankrej and the South African Africander are therefore not in as great demand as formerly for that purpose, but the cultivation of heavy land in many parts of the tropics still needs such cattle as the Ongole and the Krishna Valley.

The demand for work cattle on light land and for general farm work is maintained and, at least in Africa, is increasing as more settled forms of agriculture are practised or the holdings are increased to a size which justifies the use of animal power.

The smaller, more active types are generally favoured, chiefly because they are more thrifty and can be manoeuvred more easily in restricted space. The type of cattle suited to meet these requirements are the Indian Hallikar and Kangayam, the East African Nandi, the Butana of the Sudan , the West African Tuareg and the Sokoto Gudali.

Dairy Type

While it is possible to a certain extent to evaluate by eye the suitability of an animal for work or beef production, it is not possible to estimate its potential milk production in this way. Many tests have shown that attempts to guess the productive ability of average or above-average milking cows by their conformation are just as likely to be wrong as right, even when the guess is made by experienced stockmen.

Nevertheless, in so far as the average conformation of the cows of a breed tend towards what in the temperate zone is known as the milking type, it does indicate a suitability for milk production, although this may not be true of any particular cow of the breed.

Among the numerous characteristics ascribed to the ideal milking cow those that can with some reason be associated with production are: a placid disposition; a well formed udder, that is, with width and depth extending forward and backward rather than downward, and with a well-developed milk vein; and an overall wedge-shaped appearance of the frame resulting from a fine narrow shoulder, moderate girth, a deep belly and a broad quarter. It must however be emphasized that many of the best tropical type cows do not conform to these specifications.

While there are a few tropical breeds of cattle that could be described as of purely dairy type, such as the Sahiwal and Red Sindhi in India, most cattle used for milk production are dual or triple purpose.

Breeds such as the Tharparkar and the Hissar of India, the Kenana of the Sudan, the Nandi of Kenya and many others produce useful amounts of milk for domestic consumption. In many parts of the tropical world such as Australia, the Pacific Islands and tropical America, there are no indigenous tropical cattle and although small numbers of tropical cattle have been imported from India and Africa, milk is produced almost entirely from cows of temperate breeds.

In tropical Australia the Jersey and Shorthorn breeds are used while in tropical America, Jersey, Friesian, Brown Swiss and breeds derived from the original Spanish cattle are the most important. In

Jamaica the new breed known as the Jamaica Hope has been developed. It contains a small amount of tropical blood and appears to be one of the most productive milking breeds in the tropical world.

In the past the introduction of high-producing temperate type milking cattle into the tropics, either as purebreds or to grade up indigenous stock, was considered a panacea for the comparatively low productivity of the local dairy cattle.

Time has shown this policy to be an almost complete failure in Asia and Africa, but not in Australia and America. This may be due to very profound differences in the standard of feeding and management practices between Australia and America on the one hand and Asia and Africa on the other.

This does not mean that the productivity of dairy cattle in tropical Australia and America is satisfactory by temperate zone standards, but the productivity of many pure-bred temperate type dairy herds in these areas is as high as, if not higher than, the productivity of the best herds of tropical type cattle in India or Africa.

In the southern states of the U.S.A. an attempt to evaluate the productivity of Red Sindhi and Red Sindhi temperate type crossbreds in the sub-tropical area was partially abandoned as it was found -that neither the pure-bred nor crossbred Red Sindhi produce as well under American feeding and management practices as do the temperate type cattle.

It is probable that there is no single policy that can be applicable in all tropical countries with regard to the type of dairy cattle that should be used. In Asia and Africa it is likely that tropical breeds will give the best results, but as disease is brought under control and feeding and management practices improve, it may become practicable and economic to use temperate type dairy cattle or crossbreds in certain favoured areas.

In tropical Australia and America as has been stated above the best herds of temperate type cattle already produce better than most tropical type dairy stock and it is unlikely that tropical type dairy stock will be introduced on a large scale. These problems are discussed more fully in a later section.

REPRESENTATIVE BREEDS OF TROPICAL CATTLE

Asian (Indian)

For fuller details of Indian breeds the reader is advised to study the publications of Olver and Joshi and Phillips.

Sahiwal

This breed has proved to be the best dairy type in the Indian subcontinent and is probably the best indigenous tropical dairy breed in the world. The Red Sindhi closely resembles it, but is smaller in every respect.

The home of the breed is in Pakistan in an area situated between 29° 5' and 30° 2' latitude north where the average rainfall is 11 inches (27.9 cm.) per annum and the heat in the summer is intense; the maximum shade temperature may be as high as 118° F. (47.8° C). but the average for the hottest summer month is 107.8° F. (42.1° C.) and the average minimum winter temperature is 41.7° F. (5.4° C.).

Sahiwals are heavily built, short-legged, lethargic animals of a reddish dun or light brownish red colour, sometimes bearing white markings. The skin is loose and pliable, the hair very fine. The udder is generally large and often pendulous.

The breed has been successfully used in East Africa, and has been introduced into Jamaica for crossbreeding purposes as well as into other tropical countries where the rainfall is not too high. The bulls are very slow at service.

Ware (1941), in recording the milk yield of the cows of one of the best herds of Sahiwal in India, states that of 289 lactations, 117 exceeded 6,000 pounds (2,722 kg.), 66 more than 7,000 pounds (3,175 kg.), 30 exceeded 8,000 pounds (3,629 kg.), 16 more than 9,000 pounds (4,082 kg.), six exceeded 10,000 pounds (4,536 kg.), and one cow produced 12,000 and another 13,000 pounds (5,442 and 5,895 kg.). The average fat content of the milk was 3.7 per cent.

Gir

The breed is native to the Kathiawar peninsula of western India in an area situated between 20° 5' and 22° 6' latitude north, where the average maximum shade temperature in summer is 98.0° F. (36.7° C.) and the average minimum is 60° F. (15.6° C.) in winter. The average annual rainfall is 20-25 inches (50.8-63.5 cm.) in the northern region and up to 45 inches (114.3 cm.) in the southern.

There are several points of conformation which easily distinguish this breed from all others; they are: the pronouncedly convex, broad forehead overhanging the eyes so that it gives a heavy-eyed appearance, very long pendulous ears running to a pointed extremity near which there is a notch, peculiarly curved horns and an excessively loose skin with a large, low-hanging sheath in the males.

The colour is often mottled red of varying intensity but is more generally white with clearly defined patches of red, brown or even black; roaning is not uncommon. The cows are only moderate milkers, but selected ones are used extensively in Bombay dairies.

The bullocks are slow and if worked on hard roads need to be shod; they are powerful workers on the farm where they will do, on the average, a ten-hour day. The breed is best suited for beef production and has been much used for improving beef breeds throughout the tropics.

Ongole

This is a northern Madras breed which is reared on alluvial black cotton soil in a climate where the mean maximum shade temperature is highest in the month of May at 104.8° F. (40.4° C.) and lowest in

December and January at 64.3° F. (17.9° C.), and the annual rainfall is 30-35 inches (76.2-88.9). They are white in colour but the hump, the neck and parts of the head of the bulls are dark grey. The hide, which is fairly thick, is loose, elastic and bears fine hair.

They have a short neck, large hump and a long, broad back, features that are associated with both work and beef cattle, and it is in these respects that they excel. For tropical cattle they mature early and commence their working life as two-year-olds.

The cows are fair milkers, according to tropical standards, and produce an average of 3,030 pounds (1,374 kg.) in a lactation period of 313 days and an average intercalving period of 479 days. Ongoles, or Nellores as they are sometimes called, have been used extensively since 1895 in tropical America for improving beef breeds but they have been in great demand as work cattle throughout South-east Asia since long before that date.

Kankrej

The home of this breed is in a sandy, treeless country with a mean maximum shade temperature of 110° F. (43.3° C.) in summer and a mean minimum of 51.2° F. (10.7° C.) in January, with an average annual rainfall of 20-30 inches (50.8-76.2 cm.). Its primary function is heavy draught, but selected cows make good domestic milk producers. They are large, upstanding animals, the average bullock being about 54 inches (132.1 cm.) at the withers.

The head is carried high and bears strong, lyre-shaped horns connected with a very prominent crest. The colour is steel grey, darkening to the extremities, but at birth it is russet. The skin is fairly thick and very loose. The breed has been used quite extensively in the Americas for crossbreeding for beef production.

Hallikar

This breed is similar to several others in South-east India, such as the Kangayam, Killari and Amrit Mahal which are probably derived from it. All are grazed over what is mainly a red laterite soil in a climate where the shade temperature ranges between 60° F. (15.6° C.) and 95° F. (35° C.). The annual rainfall is on the average 31-35 inches (78.7-88.9 cm.).

In colour the Hallikar is dark grey, deepening at the extremities but lightening under the belly, neck and face. It is of moderate size and of compact muscular appearance, has a rather thin neck, a fair sized hump, a long, broad back, short, good legs and sound, small feet. The skin is fairly thick with short silky hair. The horns are typical of what is known as the Mysore type of cattle.

The breed is essentially a draught one, the bullocks being noted as fast and steady workers; they are said to do up to 40 miles (64.4 km.) a day on a rough road pulling an average load (Plate 10).

East African

The breeds are not so clearly defined as are those of other tropical regions. The reader is referred to Faulkner (1957), Epstein (1955) and Joshi, McLaughlin and Phillips (1957) for fuller information.

Nandi

These cattle are of the East African Shorthorn zebu type. They are bred chiefly in an area of Kenya where the average annual precipitation is 45 inches (114.3 cm.), the average monthly maximum shade temperature lies within the narrow range of 85°F. (29.4°C.) to 95° F. (35°C.) and the average monthly minimum is about 57°F. (13.9°C.).

The cattle are dairy type. In height they are about 45 inches (114.3 cm.) at the withers, the average weight is only about 675 pounds (306.2 kg.) for cows and 850 pounds (385.6 kg.) for bulls.

The body colour and markings are very varied but whole red with black points is possibly the most common. The head is long and the profile straight. The horns are short and fine. The hump is fairly big in bulls but often very small in cows. The dewlap is well developed. The legs are small with fine bone. The cows have a long productive life and a good milk potential, as judged by tropical breed standards.

Boran

This breed has been derived from the large zebu and in the unimproved stage is found in greatest numbers in the drier Northern Province of Kenya, Southern Ethiopia and Somaliland but they, or types very closely resembling them, are used as the basis for improved beef herds on ranches throughout East Africa.

They are noted for beef production, but they make good draught oxen and some of the cows are satisfactory milkers. They are about 48 inches (121.9 cm.) high at the withers. Bulls weigh up to 1,500 pounds (680.4 kg.) and mature, well-fed bullocks reach an average of 1,100 pounds (499 kg.).

The coat colour is very variable, but grey is the most common. The head is small, the face straight or very slightly dished, and the horns are short. The hump is fairly well developed. The general conformation quite closely resembles that of the improved beef breeds.

Kenana

This is a type of zebu favoured by the Arab pastoralists along the Blue Nile; it does not thrive further south. The average annual rainfall in the country it inhabits is about 17.5 inches (44.5 cm.) of which three-quarters fall in the months of June to September. The air temperature reaches an average monthly maximum of 113°F. (45°C.) in April and a minimum of 51°F. (10.6°C.) in January.

The colour is steel grey, darkening over the limbs, head and tail to black points. Like so many other grey zebus, the colour at birth is light brown. The hair is short and fine, the hide thin and pigmented. The hump of the bull is large but that of the cow is small. The head is long, the brow broad and the profile often convex.

The breed is of the dairy type and the cows have the general conformation associated with it, but they are slow to mature. The bullocks make good work oxen. The average weight of the yearling bulls at the Gezira Research Farm is 333 pounds (151 kg.), and that of the heifers 312 pounds (141.5 kg.). At maturity, that is at about five years old, the bulls weigh on the average 1,100 pounds (499 kg.) and the cows 884 pounds (401 kg.).

Africander

This breed is associated with regions in South Africa where the humidity is low, air temperature high, rainfall uncertain and solar radiation intense. It has been developed for work and beef production. The modern type of steer stands up to 53 inches (134.6 cm.) at the withers, has a girth of 73 inches (185.4 cm.) and a body length of 58 inches (147.3 cm.). The average live-weight at maturity is about 1,100 pounds (499 kg.) but bulls often weigh over 2,000 pounds (907 kg.).

The coat colour is dark to light brown or dark yellow with sometimes white markings on the belly. The skin bears yellow pigment. The head is characterized by a broad, convex brow, capped with a prominent frontal crest from which cream coloured horns taper downward and backward.

The ears are small and do not droop. The dewlap is big. The hump, loins and thigh are particularly well developed. The rump is adequately covered but has a slight fall to the pin bone. The legs are lean, strong and clean, the back fine and the feet small and hard.

West African

For fuller information, reference should be made to Mason (1951), Faulkner (1957) and to Joshi, McLaughlin and Phillips (1957).

Sokoto (syn. Gudali)

These are from the Province of Sokoto, Nigeria, where the average annual rainfall is 27 inches (68.6 cm.), the mean maximum shade temperature in the month of April is 106°F. (41.1°C.) and the mean minimum in January is 60°F. (15.6°C.). They are maintained, mostly under a system of transhumance, to supply the immediate domestic needs of Mohammedan tribesmen, in grazing conditions somewhat more favourable than those available to other breeds of cattle in the region.

The cattle are some 50 inches (127 cm.) high at the withers. The colour is variable, but light dun to grey and white are the most common. The hump is well developed, the dewlap and sheath large, the body compact and fleshy.

With stall feeding and under good farm conditions it has been found that there is some justification for the reputation the breed has as a dairy one. The cattle are comparatively docile and are easily trained for farm work, they are also well suited for beef cattle, not only because of their conformation but also because of their ability to fatten easily. The breed is a popular one among the farmers over a wide area of Northern Nigeria and beyond.

Red Bororo

This West African Longhorn is a breed maintained by nomads of north-eastern Nigeria and the former French Niger under the hardest conditions. The average annual rainfall in the region is about 25 inches (63.5 cm.). The mean maximum shade temperature in summer is 90°F. (32.2°C.) and in winter 67°F. (19.4°C.).

The mature cattle are, on the average, 57 inches (144.8 cm.) high at the withers. The head is carried high and bears long, lyre-shaped horns giving a general impression of great size, but the body is narrow, the legs are long and the average weight is only about 900 pounds (408.2 kg.) at maturity.

The gait is long and free, cattle of the breed are said to cover great distances at remarkable speed when under pressure. They are poor milkers and are not amenable to physical discipline but they are remarkably intelligent and readily answer the directions of a familiar voice.

White Fulani

These are bred by nomadic tribesmen of northern Nigeria under conditions somewhat less severe than those in which the Bororo exist. The climatic conditions to which the breed is accustomed are: average annual rainfall 43 inches (109.2 cm.), average maximum shade temperature in April, 98°F. (36.7°C.); average minimum temperature in December, 56°F. (13.3 C.).

The body colour is grey with black points. The average height is 51 inches (129.5 cm.) behind the hump. The neck of the males is muscular, the hump fairly large and the shoulder rather coarse. The body is compact and fleshy but rather mean at the rump. The head is small, the face straight, the horns of medium length. The skin is thick and pigmented.

N'Dama

This is one of the coastal breeds of very small cattle which are resistant to trypanosomiasis. Its origin is in former French Guinea and Senegal where the maximum shade temperature in summer is about 86°F. (30°C.) and the minimum 74°F. (23.3°C.). The average annual rainfall being 58 inches (147.3 cm.).

They are active, stocky, fine-bred looking animals with most of the characteristics of the specialized beef breeds but their average weight is only 660 pounds (299.4 kg.) and their height only 43 inches (109.2 cm.).

Elsewhere under improved conditions and after selective breeding it is reported that bulls up to 1,200 pounds (550 kg.) can be obtained. They are dun or fawn coloured but some have white markings. They have no hump and the skin, which is thin and covered with fine hair, has no folds; the dewlap is negligible and the sheath small.

Their primary use is for meat production but they make fairly good light-draught oxen. They are poor milk producers but they mature early and are exceptionally fertile. The breed has been used successfully throughout and on the fringe of the tsetse fly coastal areas of West

Africa for grading up other breeds. They have also been used in the Congo.

American

The *Santa Gertrudis* breed has been developed at the King Ranch in Texas and is a stabilized crossbred being approximately five-eighths Shorthorn and three-eighths American Brahman. These cattle, when fed and finished on grass alone, are suitable for production of beef in semi-arid sub-tropical conditions where British breeds are unable to withstand the heat, the parasites and the work entailed in collecting their food, and where the acclimatized Longhorn cattle failed to produce beef suitable for the American market.

The breed is notable for early maturity, hardiness, propensity to fatten and as an economic food converter. Under sub-tropical and tropical conditions its productivity compares favourably with the best European breeds in the temperate zone.

The body colour is red and the skin is pigmented red or black; the thin hide is loose, but not excessively so. The hump is small in the bulls and just perceptible in the cows. The head is broad, the forehead slightly convex and the profile straight. The back is broad, the trunk deep, on fleshy limbs of medium height.

The breed has been introduced to Australia, Cuba, Brazil, Fiji, South Africa and elsewhere.

The *Beefmaster* is a new unorthodox breed that has also been developed in south Texas, though the main breeding centre is now in Colorado. It is a crossbred that has been stabilized for type, but not for colour and it is approximately one-quarter Shorthorn, one-quarter Hereford and one-half American Brahman.

This new breed is an outstanding beef producer. In 1955 the average weight of the bull calves at seven months of age was 599 pounds (271.7 kg.) out of two-year-old cows and 631 pounds (286.2 kg.) out of three-year or older cows. The cows are very fertile and produce calves at two years of age off poor range country.

The *American Brahman is* a breed that has evolved in Texas from an unknown mixture of Ongole, Kankrej, Gir, Krishna Valley and perhaps some British stock. It is a large animal; the mature cows weigh 1,000-1,500 pounds (453.6-680.4 kg.) and the mature bulls 1,600-2,200 pounds (725.7-997.8 kg.).

The calves are small at birth and weigh only 50-60 pounds (22.7-27.2 kg.), but they grow rapidly. Individual animals vary in colour from light grey or red to almost black, the mature bulls usually being darker than the mature cows. The breed is characterized by a large hump, pendulous ears, a large dewlap and long legs, but the best animals have a surprisingly good beef conformation.

The American Brahman is widely used for crossing with British breeds and with Charollais in the Americas. They are being exported in large numbers all over the world.

The *Criollo* is the descendant of the original Spanish stock imported into the Americas and there are many different synonyms for this breed in the different Latin American countries. As the result of severe natural selection the descendants of the original temperate type stock are well acclimatized and are immune to tick-borne disease.

The breed is dual or triple purpose but selection is now being made for a milking type at several centres. The Criollo has short, fine, reddish yellow hair, a red or black pigmented skin that wrinkles around the eyes and the neck; its body is fairly large and angular, it has a large girth, a moderately pendulous dewlap and rather small horns. The mature cows weigh 900-1,100 pounds (408.2-499 kg.), and the adult bulls up to 1,500 pounds (680.4 kg.).

In Venezuela where a milking type is being developed the average production of one of the herds is 5,280 pounds (2,396 kg.) per lactation and with improvements in management it is considered that average yields of approximately 6,600 pounds (2,994 kg.) of milk could be obtained.

The *Jamaica Hope* has been evolved by cross-breeding Jersey and Sahiwal in Jamaica. This work commenced in 1911 and one outstanding line, the '*Norbrook*', has been produced. It is now considered that all the important genetic factors of the Jersey and the Sahiwal in Jamaica are present in the Hope breed and the herd has been closed to imported sires. The average milk production of mature Jamaica Hopes in 1955 was 9,662 pounds (4,382.6 kg.) in a 305-day lactation, the highest yield obtained so far being 20,752 pounds (9,413.9 kg.) of fat corrected milk.

DAIRY CATTLE

Milk for direct human consumption and for manufacturing milk products is obtained from many kinds of domestic livestock, but approximately 90 per cent of the world's supply is produced by various types of domestic cattle. The proportion of the total milk produced by domestic livestock other than cattle is probably higher in tropical than in temperate countries, but in tropical countries cattle undoubtedly still produce most of the milk.

That milk production in tropical countries is a particularly difficult problem is widely recognized. According to McCabe (1955) the world only produced enough milk to meet 60 per cent of the world population's minimum dietary needs in 1953, but there were surpluses in some areas and shortages in others, as countries in the temperate zone containing 40 per cent of the world's population produced 85 per cent of the world's milk.

Table 7.2: Estimated total milk production and production per capita in temperate and tropical climatic areas.

Areas	*Total milk production lb. × 10^6*	*Population × 10^6*	*Production per capita lb. (kg.)*
Temperate			
North America	138,067	174.9	789(357.9)
South America	15,123	26.9	562(254.9)
Western Europe	202,228	305.9	661(299.8)
Eastern Europe	39,257	88.4	444(201.4)
U.S.S.R.	57,025	205.0	278(126.1)
Asia	6,220	126.8	49(22.2)
New Zealand	11,748	2.0	5,874 (2,664.3)
Tropical			
Central America	1,080	9.4	115(52.2)
Caribbean	2,201	16.9	130(59.0)
South America	15,163	90.2	168(76.2)
Africa	5,191	111.7	46(20.9)
Asia (excluding India)	2,262	164.0	14(6.3)
India	42,360	367.0	115(52.2)
Oceania (Islands)	145	2.7	54(24.5)

In fact in temperate countries there was an average of more than 500 pounds (226.8 kg.) of milk per head available, while in tropical countries there was only about 60 pounds (27.2 kg.).

This is, of course, only the overall picture, as there were areas in the temperate zone with a lower and areas in the tropical zone with a higher milk production, as can be seen from Table elsewhere in this chatper and Figure elsewhere in this chapter.

These local variations are often due to local dietary and agricultural customs or to special political and social reasons, and are not necessarily due to the inherent problems of milk production. It could be argued that if the production of milk presents particularly difficult problems in the tropics, then the simplest solution would be to increase production in the temperate zone and distribute the surplus as condensed, evaporated or dried milk in tropical countries.

The fact is that, even if the economic and political problems of distributing surpluses on a world scale are solved, there are other very good reasons why a dairy industry should be encouraged in most tropical countries.

Until recently most farming in this region has been either large-scale plantation or small-scale subsistence farming. Large plantations, though in some ways efficient, are often politically and socially undesirable, and subsistence farming is usually both inefficient and socially undesirable.

In the past both systems have encouraged exploitation of soil resources, and it is now generally agreed that it is desirable to introduce some form of mixed farming. Milk production can be readily associated with mixed farming and the efficient milk producer can often obtain a relatively high income from a comparatively small farm.

Low milk production in the tropics is due to an interaction of climate, disease, breeding, feeding and management problems, and these vary in their importance from country to country and from region to region. The effect of climate on milk production is particularly complicated as climate itself is a complex of so many variable factors; and it varies widely from country to country within the tropics.

It has both a direct and an indirect effect on dairy cattle. The problems of disease are likewise complex and varied. Aspects of the effect of both climate and disease on milk production are discussed in relation to the most suitable methods of breeding, feeding and management in the tropics.

Milk and milk products are relatively expensive though very desirable foods, so that ultimately the growth of the milk industry depends on reducing the costs of production and increasing consumer purchasing power. The increasing industrialization of many tropical countries with a consequent growth of urban population and purchasing power, should create a new demand for milk and milk products.

This is a challenge to the tropical milk producer that he can meet only if he applies new knowledge of dairy husbandry to his own problems. Increased productivity will reduce the cost of milk and milk products and this will generate a still greater demand.

MANAGEMENT AND FEEDING

Systems of Management

Today, there are at least three types of dairy farmer in the tropics: the peasant or subsistence producer, the specialized dairy farmer usually of medium size but who may operate on a very small scale, and the very large scale producer who may work individually or co-operatively as a member of a union.

It is recognized that the vast majority of milk producers in the tropics are peasant farmers, who first satisfy their own requirements, and then sell any surplus that they may have either as liquid milk or as some form of milk product.

A survey conducted by the Indian Council of Agricultural Research in 1936-37 in some of the major animal-breeding tracts in India showed that the number of milk cows per holding was 4.2 in Kankrej, 4.1 in Ongole, 1.9 in Bihar, 2.0 in Hariana, 2.3 in Kosi and 3.2 in the Central Provinces.

There is no reason to believe that the results of a similar survey today would be very different, and good reason to believe that a survey in other peasant communities in the tropics would demonstrate that the situation in India in this respect is normal.

According to Singh (1950) the average annual production of cows in India is 700 pounds (317.5 kg.), and as it is known that the production of cows in the urban areas and on the larger farms is higher than this, it can be assumed that the average production of the village cow is less.

Of this small production 33 per cent is used or sold as liquid milk, 44 per cent used for making ghee and 23 per cent for making other milk products. Thus the peasant producer can only make available for the liquid milk market at most one-third of what is already a very low production.

In fact, a growing proportion of the liquid milk sold in the urban areas is produced by a small number of relatively large producers or from some form of co-operative production scheme such as the milk colonies in India.

As the standard of living gradually rises in tropical countries it is likely that there will be some reorganization of land holding, and that the production of liquid milk will become increasingly the province of specialized dairy farmers.

It is unlikely that the peasant farmer could apply all the management, feeding and breeding practices advocated, but even the smallest specialized dairy farmer should be able to apply most of them. Specialized dairy farms, unlike specialized beef holdings, do not have to be extensive in order to provide a decent standard of living for the farmer.

In fact there is intrinsic merit in the intensive dairy farm run by an owner-occupier, that must necessarily be relatively small, as one man can only handle a limited number of animals even with the most efficient labour-saving equipment. It is significant that in some of the best dairying areas in the temperate zone, such as Denmark and New Zealand, the average size of the specialized dairy farm is small.

In areas in the tropics where there are extensive dairy farms, such as northern Queensland in Australia, the standard of management is not usually very high. If milking machinery is to be installed, and this should be the eventual aim everywhere, though it would be completely uneconomic in most tropical countries at the present time, then the intensive dairy farmer should keep at least twelve dairy cows and their followers, and in the humid tropics he should be able to do this on an area of 15-20 acres (6.1-8.1 hectares).

Large-scale dairy farms can of course be intensively managed. There are a small number of these organizations in tropical countries that are privately owned, and there is a movement, particularly in India, to group small producers into large units, where the individual owns and manages the animals while a collective management provides communal services such as housing, feed, and machinery, and also purchases, processes and markets the milk.

Cows on intensive holdings can be managed indoors or outdoors or partly indoors and partly outdoors. The relative merits of the two systems will be discussed in a later section.

The subsistence or peasant producer

Peasant producers normally never keep animals solely for milk production. A cow is expected to work and milk during her lifetime and when she is too old for work purposes, if social custom allows, she will be sold for slaughter.

Feeding will be haphazard and management poor. Peasant farmers do not generally understand the value of pasture nor do they farm sufficient land to enable them to grow special forage crops for their animals. Cows are expected to graze or browse along the roadsides or on stubbles or they are tethered and fed cut forage. They may receive some offal concentrates such as rice bran.

Any producer is however usually responsive to advice if he considers that it is sound and is convinced that it will improve his income. He can be taught the importance of good feeding and shown how to grow tree crops that will produce browse in the odd corners of his holding or along paths or paddy field bunds; he could be convinced that it is a bad practice to tether cows out in the open on hot days; and he can be shown that his cows will let down their milk without the calf suckling at the same time, and be persuaded that this is a desirable practice.

Peasant farmers cannot be considered as milk producers within the normal meaning of the term, though it is recognized that they are an important factor when assessing the total production of milk in the tropics, and if a suitable organization for the collection and processing of liquid milk is available, as it now is in some parts of India, the very small quantities of milk that the peasant farmer produces can be channelled to the urban consumer.

In India, for example, the Bombay Government has set up an organization in Poona that collects 6,000 pounds (2,722 kg.) of milk daily from 357 producers in 44 villages and another in Hubli-Dharwar that collects 2,000 pounds (908 kg.) daily from 1,000 producers in 51 villages.

A similar example of the incentive to milk production afforded by a suitable organization for collection and processing is that from Jamaica (1955) where 1,896 suppliers sent 1,530,795 gallons (6,958,994 litres) of milk to the condensery during the period January to August, 1956. There were 6,430 farms of from one to 24 acres (0.405 to 9.7 hectares) in size supporting dairy cattle and 16,550 farms of similar size on which dual purpose cattle were maintained.

It is estimated that at the present rate of progress 15 million gallons (68 million litres) of milk may be handled annually within another ten years.

The specialized dairy farmer

The specialized dairy farmer is potentially the most important milk producer in the tropics. There are today many peasant producers who could rapidly become small specialist producers, but it is unlikely that the larger specialist producer, common in some dairying areas of the temperate zone, will become an important factor in milk production in the tropics for a considerable time, if at all.

If the evolution of peasant producers into small specialist dairy producers is to proceed rapidly it is essential for the authorities in each country to provide them with organized marketing facilities and efficient advisory and other auxiliary services.

Central purchasing and milk processing plants are required, together with advice on forage production, feeding and dairy hygiene, and loans for the purchase of the necessary equipment, buildings and stock.

The provision of an artificial-insemination service can be most helpful to the small specialist producer. It enables him to improve his breeding stock without the necessity of purchasing expensive bulls and allows him to keep an extra cow instead of a bull. The latter will increase his total production by a considerable fraction if he only keeps a small number of cows.

The large-scale producer

There are a small number of large-scale private or government-owned dairy farms in most tropical countries. In India, for example, there are many large government dairy farms, about forty military dairy farms producing milk for the armed forces and some large-scale private farms, while in Singapore there is a large-scale privately owned farm, and in the Panama Canal Zone a large government-owned farm.

Management, feeding and breeding practices are usually superior on these large farms to what they are on the smaller farms, not because they are inherently more efficient units, but because of the superior knowledge and resources of the operators. It is unusual for the management of these large units to be of the same standard as that of the best dairy producers in the temperate zone.

The majority of the large producers manage their stock indoors, either in sheds or in yards. Some of them breed their own replacements, indeed many of them are breeding centres for the country in which they are situated, but others buy in milkers. Where replacements are bought in, the culling rate is high, sometimes as high as 30 to 50 per cent per annum.

Some of the most efficient large-scale producers operating on this system are to be found in Los Angeles County in the United States. This area is not in the tropics but the climate is very hot in the summer and conditions are quite similar to what they are in parts of the arid tropics. The average size of the herds in Los Angeles

County is 250 milking cows. The amount of milk yielded by 20,000 cows is recorded, the average individual production in 1956 being 13,120 pounds (5,951 kg.) containing 496.5 pounds (225.2 kg.) of butterfat. Pipeline milking machines and automatic feeding devices are used but are not universal.

Almost all feed is bought in, the cows being fed alfalfa hay, molasses and grain. The culling rate is close to 50 per cent per annum.

In India there has been a very interesting development of large-scale production. The best-known scheme is the Aarey Milk Colony in Bombay. This city has a population of approximately 2.5 million people and 60,000 milk cows and buffaloes were kept within the city or suburbs under most unhygienic conditions by some 3,000 milk producers.

These producers have been directed to bring their small herds together in one area some 20 miles north of Bombay city, where an authority provides housing, facilities for milking, manure collection and utilization, feeding, watering and fodder production, together with breeding and other advisory services.

The individual producer manages, feeds and milks the animals himself. In 1953 the Aarey Milk Colony was distributing 30,000 gallons (136,380 litres) of milk daily and in addition was processing and distributing milk collected at Anand some 260 miles away from Bombay.

The advantages of the Milk Colonies are that they are able to provide the small farmer with modern buildings and equipment. They are also able to control the adulteration of milk, which is very common in the tropics, can purchase feed in bulk, advance money on the security of the milk that they receive and provide advice in a manner that would be impossible if the producers were not all sited in one area.

The Calf

The first decision that a farmer has to make is what calves he will rear. A breeder will raise all his heifers and bull calves, and sell the surplus as breeding stock. Commercial milk producers have a choice of several policies. If they raise single-purpose dairy animals such as Jersey, Sahiwal or Jamaica Hope they will want to rear all their heifers but may wish to dispose of their surplus bull calves at or soon after birth.

In the temperate zone these calves would be sent to a factory where they would be slaughtered and the meat used for manufacturing purposes. In most tropical countries it is not possible to do this, but the calves can be slaughtered on the farm and the meat cooked and eaten or used for the feeding of stock such as pigs and poultry; or what is more likely there may be a local market for surplus, live, young calves.

If the farmer raises dual purpose animals, and this is more usual in the tropics as most tropical breeds are dual or triple purpose, he can keep all his surplus bull calves, castrate, raise and sell them as steers; either for meat or for work purposes.

There is also a third policy but it is not widely practised in the tropics. The farmer can keep a beef or work type bull and use him to serve the poorest producers in the herd, only using the better cows for breeding future dairy replacements. The crossbred calves, both heifers and bulls can then be raised for beef or work.

This is a very desirable policy but it cannot usually be practised because calf mortality is high and the farmer needs all the dairy heifers that he can rear for replacements in his dairy herd.

Calf-rearing systems

Calves can be reared indoors or outdoors or partly indoors and partly outdoors. In cold temperate countries calves are often reared indoors for the first six to nine months of life because of lack of grazing and the cold, but in warm temperate countries such as New Zealand calves are usually reared out of doors all through the year.

In the humid tropics, where temperatures are high all the year round and grass growth is more or less continuous, it might be thought that it would be advantageous to follow New Zealand practice. This is not so, however, for three reasons.

First, it is difficult to keep grass in the calf paddocks in a young, palatable stage; tropical grasses grow and mature very quickly and at maturity have a high fibre and low protein and mineral content, and are not suitable for calf grazing.

Secondly, a humid tropical environment is ideal for internal parasites and it is very difficult to keep the calves free from massive infection if they are grazing.

Thirdly, even if the calf grazings are well shaded the calf is exposed to considerable climatic stress, at a stage of life when it is least resistant. The combination of innutritious forage, parasites and climatic stress is inimical to calf life and calves raised under these conditions usually have a high mortality rate.

In the arid tropics, the risk of parasite infection is less than in the humid tropics but grass growth is very seasonal and during long periods of the year no suitable grazing is available for calves.

In both the humid and arid tropics calves are best reared indoors, though in the arid tropics they may be reared outdoors during the period of grass growth.

The most practical indoor rearing system is one in which the calves are reared in a calf shed sited in an earth or gravel yard where they ma exercise. The essentials of a good calf house are that it should pride adequate shelter, should be easily cleaned, and should keep the young calves warm in cool weather and cool in warm weather. It may be no more than a simple roof over a concrete base on a well-drained site. Side walls are not necessary in the humid tropics as long as the roof has a wide overhang to keep out driving rain. Under the roof provision should be made for individual feeding, separate pens, etc. Tubular steel is the most suitable material from which to construct individual calf ties.

The yard must be kept free from all vegetation in order to reduce the chances of parasite infection. The calves should have access to water, preferably under the shade of the roof, concentrates, good quality forage and minerals. They should have a dry bed on which to lie, and are best bedded down on straw, but this is not essential. It is usual to raise and feed the calves individually for the first few weeks of life and after that in batches.

The internal layout of the type of calf shed described can easily be altered to suit particular circumstances. In some tropical countries individual calves are raised for the first few weeks of life in a stall with a slatted floor; through which the farces and urine drop. This is a good system but more costly and not essential if calf pens are cleaned daily and kept dry.

Calves that are running in batches often suckle or lick each other after feeding and it is a good practice to keep them in their ties for some time after milk feeding. Hair swallowed by the calves after suckling each other forms hard balls in the abomasum and this is a constant cause of digestive disturbance.

If for any reason dairy calves have to be reared outdoors in the tropics, they should not be reared in special calf paddocks but should be rotated around the cow grazings. It has been found in New Zealand that the best system is to rotate the calves around the grazings at least ten days ahead of the dairy herd.

In order to practise this the farmer needs at least fourteen to twenty separate paddocks. Whether calves are reared indoors or outdoors it is usual to bring them to one central point to feed milk and concentrates, as it is easier to walk the calves to the food than to bring the food to the calves.

Feeding

The digestive tract of a young calf differs from that of a mature cow, and the calf does not function as a ruminant until it is a few weeks old. In the calf the capacity of the true stomach or abomasum is 70 per cent of all four stomachs, whereas in the mature cow it is only seven per cent. When the calf suckles, the milk passes directly into the true stomach or abomasum, and only if the calf drinks too much does any milk pass into the rumen. Milk going into the rumen in a small calf may curdle, and then, because rumination has not yet started, putrefy, causing digestive disturbances. Thus it is better practice to feed the calf small quantities of milk at frequent intervals, than large amounts at infrequent intervals.

Suckled calves running with their mother are least likely to suffer from digestive disturbances as the calf suckles frequently. This is the method by which beef calves are raised but it is not practical or economic to raise dairy calves in this way. The majority of peasant dairy farmers in the tropics suckle the calf while they are milking the cow, and cows accustomed to this practice will not usually let down their milk unless they are suckled.

This is an undesirable practice as it is both uneconomic and unhygienic. The practice creates a vicious circle; the farmer starts the heifer off in this way when she is first milked and then says that his cows will not let down their milk unless they are suckled. Dairy calves can be suckled by raising them on a nurse cow; one, two or three calves being raised according to the yield of the cow.

This is quite a common practice among breeders of dairy cattle and may be a desirable practice in the tropics where calf mortality is high and general feeding and management poor. Old cows and cows with 'blind' quarters may be used for this purpose.

If the calf is not suckled but bucket fed, and this is the normal practice in dairy farming countries, then it is better to bucket feed three times a day. Many farmers bucket feed through a nipple as this slows down the rate of feeding.

The disadvantage of bucket feeding through a nipple in the tropics is that the practice requires more equipment, and that all the equipment has to be kept very clean, thus adding to the expense and the difficulties of management.

Whether the calf is to be suckled or bucket fed it is essential that it should receive the colostrum or first milk from its mother. It should be fed its mother's colostrum for three days, but if it is to be bucket fed it should only be allowed to suckle its mother for 24 hours, and after that her colostrum should be fed from the pail.

Colostrum contains twice as much dry matter as milk. Its protein content is 18 per cent as against three to four per cent in milk and is of a different quality. It contains very much larger quantities of vitamins and minerals. It also contains *antibodies* needed by the growing calf.

These assist the young calf to protect itself against disease. It is particularly important that the young calf should get colostrum within the first 24 hours after birth as its digestive tract can absorb the antibodies during this period, whereas later they cannot be absorbed. Surplus colostrum may be fed to the older calves. It is usual to mix it with milk, but this is not essential.

It is sometimes difficult to start a calf drinking from the bucket. The usual method is to starve it for a few hours and then place a hand in its mouth when feeding it so that the colostrum laps against the mouth and the calf sucks it up through the fingers, which are then gently withdrawn.

The young calf often 'bunts' while being bucket fed; this is quite normal as that is how it starts the 'let down' of milk in its mother; but it means that the feeding bucket must be secured in some way otherwise the calf will spill its food.

The calf cannot use average quality tropical roughage extensively until it is four or five months old and if it is fed solely on roughage its growth will be slow. Consequently it is desirable that calves should be suckled or bucket fed for many months and also fed concentrates and as high a quality roughage as is obtainable.

Suckled calves if raised outdoors may run out on grazing and select what grass they need but they are usually fed concentrates and cut forage indoors. Bucket fed calves are usually fed whole-milk for a time, then skim-milk and finally they are weaned when about six months old.

During this period they should also have access to grazing or be fed concentrates and cut forage indoors. The 100pound (45.4 kg.) calf requires food with a nutritive ratio of about 1 : 4 and as it grows it can gradually deal with foods of a higher nutritive ratio.

Early weaning is now practised extensively in temqerate zone countries. Calves are introduced to mealed or pelleted concentrates when they are ten days old and to good quality roughage when three weeks old.

They are weaned either abruptly when 24 days old, after being fed 6 lb. (2.7 kg.) whole milk daily, or gradually when they are given 6 lb. (2.7 kg.) till ten days old and thereafter 1 lb. (0.5 kg.) less each week until the milk is stopped when the calf is 31 days old. When thus weaned their growth is severely checked but they recover satisfactorily in due course and suffer no permanent damage.

It is necessary to compromise between the needs of the calf and the costs of rearing. In the tropics where conditions are less favourable to the calf it is probably more economic to feed the calf additional whole milk, than to attempt to raise it on the minimum quantity.

It is usual to ration the calf to eight to ten per cent of its own body weight of milk a day. Some authorities advocate the feeding of up to fifteen per cent of the body weight of milk daily, but this is not general and certainly should not be practised in the tropics where calves usually grow more slowly than they do in the temperate zone.

The milk should be warmed to body temperature and the calf should be fed three times daily during the first week, if that is practicable. In Table elsewhere in this chapter are shown details of two rations, one in which the calf is fed whole-milk for one month and one in which it is fed whole-milk for two months. It will be noted that there is no abrupt change from whole to skim-milk feeding.

Table 7.3: Whole and skim-milk feeding of calves.

Daily amounts for calves weighing 45 to 60 pounds (20.4 to 27.2 kg.) at birth. Heavier calves should receive correspondingly larger amounts.

	Ration 1		*Ration 2*	
Weeks	**Whole-milk lb. (kg.)**	**Skim-milk lb. (kg.)**	**Whole-milk lb. (kg.)**	**Skim-milk lb. (kg.)**
1	4(1.8)		4(1.8)	
2	5(2.3)		5 (2.3)	
3	7(3.2)		7(3.2)	
4	6(2.7)	2(0.9)	8(3.6)	
5	4(1.8)	4(1.8)	8(3.6)	
6	2(0.9)	6(2.7)	9(4.1)	
7		10(4.5)	8(3.6)	2(0.9)
8		11(5.0)	6(2.7)	5 (2.3)
9		12(5.4)	4(1.8)	8(3.6)
10		12(5.4)	2(0.9)	10(4.5)
12		12(5.4)		12(5.4)
14		10(4.5)		10(4.5)
18		8(3.6)		8(3.6)
20		6(2.7)		6(2.7)
22		4(1.8)		4(1.8)
24		2(0.9)		2(0.9)
26		Nil		Nil

Some farmers dilute the milk that they feed to their calves but this is not necessary. On farms that sell liquid milk, fresh separated milk can be replaced by separated milk powder or buttermilk powder. The use of milk powder for calf feeding will depend on local economic factors.

Although milk is essential for young calves they must not be fed only on milk. Calves fed in that way suffer eventually from magnesium deficiency and die.

Calves from two to three weeks of age should have access to concentrates and to the best quality forage available. The type of concentrates that can be fed obviously depends upon what is available locally. A suitable mixture is:

Coconut meal	50 parts
Groundnut meal	25 parts
Maize meal	25 parts

Warner (1951) recommends the following mixture for calf rearing in India:

Wheat bran	30 parts
Linseed cake	10 parts
Barley meal	20 parts
Jowar meal	20 parts
Maize meal	20 parts

At two months a calf will eat up to one pound (0.45 kg.) of concentrates per day, while at six months it will eat up to three pounds (1.4 kg.) per day.

When calves are fed indoors special precautions must be taken to see that they receive adequate amounts of minerals and vitamins. It is best to add a little fish-liver oil to the diet and to make available a mineral mixture.

The mineral mixture can be fed with the concentrates but it is a better practice to feed it separately. Calves should always have access to water. It is a mistake to assume that they do not require water because they are being fed large amounts of fresh or separated milk.

Special management considerations

When a dairy cow is about to calf it should be brought to a place where it can be kept under observation. Some farmers provide special calving pens but this is not essential unless contagious disease is endemic. Occasionally a cow needs assistance at calving, especially when a tropical type cow is delivering a calf got by a temperate type bull, but help should be given only when it is absolutely essential. After birth the cow will clean the calf by licking it.

Most breeds of dairy cattle are horned and horns are a disadvantage in dairy operations, though the farmer may wish to retain the horns on bull calves to be raised for work or on dual or triple-purpose heifer calves.

Dehorning is best done by cauterizing the horn bud when the calf is a week or two old either by rubbing the bud with a caustic stick till it is near bleeding, or by the use of a cylindrical redhot iron pressed for a second or two on the rim of the horn bud.

When caustic is used care must be taken that the calf does not carry it to the cow, in suckling, for instance, and that the caustic is not spread from the site of the operation, as may occur if the calf is exposed to rain soon after application.

Castration can be carried out with a knife, with crushers, or with a rubber ring. The rubber ring method is the most suitable for calves up to three or four weeks old. When the operation is performed by this method on calves when they are about ten days old it causes a minimum of pain or discomfort. The Burdizzo bloodless castrator makes the operation a comparatively safe one at all ages.

Calf mortality and disease

Calf mortality in the tropics is undoubtedly very high: often it is as high as 50 per cent but this almost invariably denotes bad management. In some areas a high death rate is partly due to the fact that unacclimatized temperate or crossbred stock are used for dairying and climatic stress is added to the other hazards of calf life, but it is mainly due to poor feeding and faulty management.

The best way to reduce calf mortality is to practise good husbandry. Some common ailments can be treated by the farmer but once the calf is seriously ill or the sick ratio is high expert veterinary advice should be sought.

The most common disease of calves is '*scouring*'. The usual symptoms are a dull appearance of the eyes, listlessness, diarrhoea and sometimes an abnormal rise in temperature and respiration rate. Scouring may be due to purely mechanical causes, nutritive causes or to infection.

Calves that drink too rapidly, or are fed cold milk, or have hair balls in their abomasum, often scour. Treatment consists of starving the calf for 24 hours and then feeding small quantities of milk diluted with a little water at body temperature.

Calves suffering from this type of scour usually recover if treated properly. Scours caused by infection are usually more serious both because there is greater danger of death to the infected individual and because usually a large proportion of the calves are infected.

The feces are often a chalky white colour and have a very offensive smell. Infection may enter through the mouth, but sometimes enters through the navel cord. Often the calves die within a very short time. The use of antibiotics in the feed of calves undoubtedly reduces the incidence of scours caused by infection.

Pneumonia often occurs when a calf is suffering from scours. The symptoms are as follows : the temperature rises rapidly, the respiration rate increases and the calf '*gasps*' for breath. The primary cause is almost invariably a virus or bacterial infection predisposed by sudden changes in the environmental temperature and humidity.

The young calf is particularly vulnerable to attack by internal parasites that thrive in the tropics. Calves suffer from the attacks of roundworms, threadworms, hookworms, lungworms and tapeworms. Symptoms of parasite attack are 'a falling back in condition', a dry rough coat, scours, a pot belly and a swelling under the jaw.

Internal parasites are best controlled by cleanliness in the calf house and by rearing the young calves indoors. The subject is discussed in other chapter of this book. If calves are kept outdoors it may be advisable to drench them monthly with a vermifuge such as phenothiazine, the dose being as follows:

Age: months	*Dose: ounces (gm.)*
2—4	½(14)
4—6	½(21)
6—12	1 (28)
12—18	1½ (42)
Over 18	2 (57)

It is usual to stop dosing once the animals are over 18 months old. Phenothiazine is effective against roundworms and threadworms and, in repeated doses, against hookworms but it is not effective against tapeworms.

If the calves are infected with tapeworms the use of phenothiazine should be alternated with the use of another vermifuge such as nicotine sulphate which is effective against tapeworm.

Calves that have been dosed with phenothiazine sometimes appear to go blind: they should be kept indoors and will soon recover. Very young calves should never be drenched.

The Heifer

The rearing of the heifer from weaning at six months of age to first calving falls naturally into two stages, one from weaning to first service, and one from first service to calving.

The aim in rearing heifers should be to achieve the maximum growth and development, and the earliest maturity consistent with cost. This ensures that maintenance costs are kept at a minimum, that there is an early return on the investment in the original animal, and that the heifer produces well during her first lactation.

There is always a check to growth at weaning. If the heifer is being rotationally grazed on good pasture the check will not be very great, but if she is raised indoors and suddenly placed out on a very poor pasture the check will be very severe.

It is now well established that the retarded growth resulting from suboptimal feeding alone has no permanent effect on future production, that is, the ultimate weight and size attained or in total milk yield. Cheap concentrates such as coconut or palm kernel meal fed in as small quantities as one half to one pound (0.23 to 0.45 kg.) per head per day to heifers 6 to 10 or 12 months of age on medium quality roughage will materially hasten gain and maturity.

Like the calf, the heifer can be reared indoors or outdoors. As indoor rearing is so expensive, heifers are usually not housed, though in some parts of the tropics they are kept indoors until they are 9 to 12 months old.

In some special areas heifers have to be reared almost entirely indoors; such as in those parts of Africa where tsetse fly is particularly prevalent; or in parts of Central and South America where the animals are liable to be bitten by vampire bats and must be housed at night.

There is no doubt that in the tropics, temperate and tropical type stock do not grow at the rate that temperate type stock do in the temperate zone. In New Zealand it has been shown that under outdoor conditions growth depends very largely upon grazing management.

Although the effect of the climatic environment is important in the tropics, grazing management practice is still a major factor in the growth of heifers. They can be either set-stocked or rotationally grazed. Rotational grazing is rarely practised.

In New Zealand experiments, rotationally grazed heifers weighed 150 pounds (68 kg.) more than set-stocked heifers at 20 months of age, and did not have to be drenched with phenothiazine, though the set-stocked heifers had to be drenched at three-weekly intervals. One reason why set-stocked heifers do not thrive as well as those rotationally grazed is that they have to work harder in getting their food. This was well demonstrated in the New Zealand experiments where the rotated heifers grazed for 7.5 hours each day, taking 48 bites per minute, compared with the set-stocked heifers 9.0 hours of grazing and 35 bites per minute.

This is even more important in the tropics than in the temperate zone as any increase in muscular work increases the heat load on the animal with a consequent depressing effect on liveweight gain.

There is no doubt that part of the poor growth of heifers in the tropics comes from set-stocking them on poor grazing at an early age, though some part is due to the effect of the climatic environment. Heifers that are outdoors should be managed as follows.

They should be shifted daily if this is possible and never grazed on any one field for more than five days. They should always be rotationally grazed around paddocks containing grass of milk-producing quality. If cow paddocks are used they should graze seven to ten days before the milking herd.

A minimum of fourteen to twenty paddocks should be provided in order to fully exploit rotational grazing techniques. Heifers raised outdoors should, whenever possible, be grazed in shady paddocks and adequate supplies of cool water should be available. Concentrates and minerals can be fed at a central point or in the field. If fed in the field they should be fed from troughs protected from rain.

Heifers that are raised indoors should be provided with good quality roughage, either freshly cut or preserved, a small quantity of concentrates, access to a mineral mix and fresh, cool water. Heifers are best raised in yards with adequate shade.

The shade should be orientated east and west, should be ten feet (3.0 metres) high and is best constructed of some form of thatch, though corrugated iron painted white or silver on top and black underneath is a suitable material.

Heifers raised indoors can be fed small quantities of cheap roughages such as pineapple pulp or bran, rice bran, citrus pulp, molasses, etc., but they must also receive adequate quantities of high quality forage.

Most tropical dairy stock and temperate type stock in the tropics grow so slowly that sexual and conformation maturity occur together, while temperate type stock in the temperate zone are sexually mature long before they are physically mature.

In the temperate zone heifers from small dairy breeds are first served at about 15 months of age while those of the larger breeds are first bred at about 18 months. Most heifers in the tropics are too small to be bred at these ages and it is better to use live-weight rather than age as a criterion of when they should first be served.

In India it is usual to breed Red Sindhi heifers when they weigh approximately 450-500 pounds (204.1226.8 kg.) and this is a suitable weight for most of the smaller breeds. Heifers of the larger breeds should be bred when they are 650-700 pounds (294.8-317.5 kg.) live-weight.

This means that dairy heifers in the tropics calve down for the first time at a much later age than is normal in the temperate zone: they are generally about three years old before so doing.

After first conception the heifer not only has to continue to grow but bear a viable calf and produce milk nine months later, so she needs to be well fed, particularly during the pre-calving period. It is advisable to bring her into the milking herd at least two months before calving.

This enables her to be fed a little extra concentrates in the milking bail and accustoms her to handling in the milking stalls. Heavy feeding a few days before calving is known as 'steaming up' and if the heifer is of a high-producing breed four pounds (1.8 kg.) of concentrates per day should be given. The concentrate mixture should be the same as that fed to the milking cows.

Table 7.4 : Standard nutrient requirements for growing heifers.

Live-weight lb. (kg.)	*Total dry matter lb. (kg.)*	*Digestible crude protein lb. (kg.)*		*Total digestible nutrients lb. (kg.)*	
		Minimum	*Good*	*Minimum*	*Good*
250(113.4)	5.9-6.9 (2.7-3.1)	0.61(0.28)	0.71(0.32)	4.1(1.9)	4.8(2.2)
300(136.1)	7.2-8.0(3.3-3.6)	0.67(0.30)	0.78(0.35)	4.9(2.2)	5.5(2.5)
400(181-4)	9.0-10.0(4.1-4.5)	0.80(0.36)	0.90(0.41)	6.1(2.8)	6.6(3.0)
500(226-8)	10.6-11.8 (4.8-5.3)	0.87(0.39)	0.98(0.44)	6.9(3.1)	7.7(3.5)
600(272-2)	12.0-13.6 (5.4-6.2)	0.94(0.43)	1.06(0.48)	7.7(3.5)	8.7(3.9)
700(317.5)	13.4-15.5(6.1-7.0)	1.00(0.45)	1.13(0.51)	8.4(3.8)	9.7(4.4)
800(362.9)	14.8-17.4(6.7-7.9)	1.06(0.48)	1.20(0.54)	9.1(4.1)	10.7(4.9)

The major diseases during this period are caused by parasites, and the general principles of their control have already been discussed. All heifers should be vaccinated against contagious abortion when they are between four and eight months of age. In rinderpest areas they should also be inoculated against this disease, when approximately eight months old.

The Milking Heifer and Cow

Management and feeding during breeding

The average gestation period of temperate type cows is 281 days, with a normal variation between 275 and 287 days in the temperate and tropical zones. Some tropical breeds have longer or shorter average gestation periods.

In temperate type cows heat periods recur on the average at 21-day intervals with a normal variation between 16 and 26 days both in the tropics and the temperate zone. In the majority of tropical breeds heat periods recur on the average at 21-24 day intervals and are often of very short duration.

In the temperate zone it is normal to advise that cows should be bred to conceive between 75 to 110 days after calving. Breeding on or about the 85th day results in calvings at 12-month intervals. In the tropics it is generally considered that it is better to breed earlier than this, and that it is more difficult to get cows in calf if they are not served before the 85th day after calving.

It is thought that they should be served at the heat period that occurs approximately 50 days after calving. It is often difficult to get the cow or heifer in calf at the first service. In the temperate zone the chances of a successful first service are about 1 : 2.3, and there is some evidence that it may be more difficult in the tropics.

This may be because the heat period there is on the average of shorter duration in both temperate and tropical type cattle. A percentage of heifers and cows are of course always sterile. In the temperate zone sterility in cows up to 10 years of age varies from three to five per cent, but when they are over 10 years old the percentage sterility rises rapidly.

There is no evidence that the percentage of sterile animals is markedly different in the tropics where dairy cows and heifers come into heat at all seasons.

It is normal both in the tropics and in the temperate zone to run a bull with the heifers, so it is important that groups of young stock should be of approximately the same weight and size. In the tropics it is often the practice to run the bull with the dairy herd, but this is not very desirable if the farmer is breeding his own replacements as it is difficult to know when the cow has been served and consequently when it will calve. It is more satisfactory to keep the bull close to the milking shed and to take cows that are on heat to him.

If a cow does not have a dry period between lactations its subsequent yield will be reduced. In temperate breeds raised in the temperate zone heifers usually produce about 70 to 77 per cent of the milk that they will produce when mature, and they reach a maximum at the fourth or fifth lactation, production declining after the seventh or eighth lactation.

Tropical breeds often produce at a maximum as heifers but the majority do so at their third or fourth lactation although that may be only 10 per cent above the first lactation yield. The reasons for this are not yet fully understood.

If the cow is calving every 12 months it should be milked for 10 months, or for a 305-day period, and rested for two months. The dry period allows the mammary gland to rest and recuperate and the cow to build up a body reserve ready for the next lactation.

A heifer should get a longer dry period than a cow, and it is advisable to dry her off after nine months so that she has a 90-100-day dry period. Cows in poor condition should also be allowed a longer dry period and undersized heifers should be given a particularly long dry period.

A short lactation is characteristic of both temperate and tropical dairy stock in the tropics so that usually the problem that confronts the farmer is that the dry period is too long.

The dry cow should be well but not lavishly fed. She should not require any concentrates until she is '*steamed up*' before calving and can be grazed on medium to poor quality roughage or on stubbles. Climatic stress will be at a minimum at this period of her life so that she does not have to be so carefully managed as a milking cow.

A cow or a heifer about to calve should be placed in a pen or in a separate paddock so that she can be kept under observation. She should not be interfered with unless she is obviously in trouble. At calving the heifer or cow loses 8 to 12 per cent of her body weight.

While calving the cow or heifer should not be overfed. A small quantity of a laxative concentrate such as rice bran is a useful feed at this time.

After calving, concentrate feeding, if not already begun, can be started immediately and it is usual to reach a peak some three weeks later. The cow should always be fed a little more concentrates than her milk yield justifies in order to encourage milk production to rise during the first part of the lactation.

Management and feeding during milking

General management

The aim of good management of dairy stock in the tropics should be to take all economically justifiable measures that will decrease the total 'heat load' of the animal or help to spread the 'heat load' more evenly over the 24 hours. As temperate zone methods of management are not based on this premise they are not necessarily suited to the tropics.

In the light of present knowledge management practices that would appear to reduce the 'heat load' or spread it more evenly over the 24 hours are: the provision of rations that do not exceed optimal requirements, particularly with regard to protein and fibre; grazing at night and not during the day, and concentrate feeding during the early morning and late afternoon; the provision of an adequate water supply both in the yards and on the grazings; the provision of natural or constructed shelter both in the yards and on the grazings; the choice of suitable materials for shelter construction; the siting of yards, bails and shelters so that they are open to the prevailing winds; the clipping of the coat of animals that are to be sheltered from solar radiation; the breeding of animals to calve down during the coolest season if seasonal milk production is not economically undesirable; and finally the provision of water sprays, forced air fans and a cooled water supply in the yards, bails or shelters, if the use of these practices can be economically justified.

Cows, like calves and heifers, can be managed indoors or outdoors. It is obvious that the 'heat load' on the animals can be more easily decreased if animals are managed indoors. In addition to the above factors the 'heat load' due to muscular work is decreased if the animals are fed indoors as forage is brought to the animals and they do not have to spend muscular energy in seeking for it.

The available information on this subject has been reviewed by Payne. The decision to adopt an indoor or an outdoor system of management is of course primarily an economic one. Will the increased milk production obtained by ameliorating conditions for the dairy cow pay for the additional cost of the buildings required and the cost of bringing the feed to the animals and carting the waste products away? Unfortunately at the present time there is no objective information on this fundamental question.

Whether managed indoors or outdoors cows exhibit very definite behaviour patterns. They show a well-established herd order and are very quick to associate pleasant or unpleasant experiences with places or persons. They should always be handled quietly and if possible by the same person. They should also be handled according to a strict routine.

Cows managed indoors develop long toes. These should be periodically trimmed. This also happens if they are grazing and walking over very soft, muddy fields as they often are in the humid tropics. Some heifers that are managed indoors develop the habit of sucking each other and odd individuals will 'suckle themselves.

If this is not prevented they will continue to do it when they are mature. A small chain attached to a ring or two rings in the nose will prevent it.

The habit of kicking is usually developed when the heifer is first milked and is due to faulty management. It is almost impossible to break the habit once it has been acquired, and as kickers disturb the milking shed routine they should be disposed of. Some cows become vicious, usually due to poor management.

In the early stages the habit may be broken by a few sharp blows on the nose, but if the habit continues the animal should be sold.

It is a good practice to groom animals that have dirt caked on them (but not just before or during milking) as this improves cleanliness at milking, but there is no evidence that it improves the health of the animals. It is a desirable practice if the farmer has time to do it, as in addition to cleaning the animal it accustoms her to handling.

Mastitis is common in milking-cows both in the temperate and tropical zones and is due to a variety of causes. One of the principal causes is bad management during milking and this is discussed in a later section. A cow suffering from mastitis should be milked last so that there is no danger of the milker carrying infection from cow to cow.

Milking technique

The object of milking is to obtain all the milk from the udder. If milking is incomplete there is a tendency for the cow to dry off too soon. The first essential, particularly when milking a heifer for the first time, is to prevent the animal getting excited or frightened.

If a heifer is to be machine milked it is important that the cups of the machine are never associated with pain or fright, because if they are, she will probably never let her milk down properly and may become a 'kicker'. The second essential is to milk quickly. This applies whether the animal is hand or machine milked.

The 'let down' of milk is controlled by a combination of nervous and hormonal action. The cow first needs a stimulus. When a calf is suckling the pull of the calf on the teat is the stimulus. The nervous system relays a message to the posterior pituitary gland which then releases a hormone known as oxytocin; this circulates for some four minutes in the blood, is carried to the udder and there induces the let down of the milk.

As the hormone is in circulation for such a short time, the quicker milking is carried out the more efficient it will be. The maximum rate of *flow* of the milk is a highly hereditary character differing in each individual. When the calf is not suckling there are many stimuli that may serve to condition the milk let-down reflex, and the establishment of a strong let-down reflex associated with the preliminaries of milking is essential for fast milking.

The idea is to produce a single let-down response at such a level that, with normal milking, all the milk can be obtained without inducing a second let down.

Milking is accomplished by the application of force to the muscles surrounding the teat meatus so that the orifice of the teat is opened and the milk discharged. In hand milking the upper part of the teat cistern is closed tightly by the hand to prevent milk flowing back into the udder cistern while pressure is applied to the milk-filled lower portion of the teat. Increased pressure inside the teat stretches the sphincter muscles and the teat meatus is forced open.

In machine milking the pressure on the inside of the teat is unchanged. The pressure inside the teat cup is lowered so that a vacuum is formed and the normal pressure inside the teat in relation to the pressure outside becomes large enough to force the meatus open and allow the milk to *flow* out.

If milk does not flow the vacuum extends inside the teat causing damage to the mammary tissues, so that the teat cups should be taken off as soon as milk flow ceases.

If the cows are milked by hand, it is better to milk with dry rather than wet hands, as wet-hand milking is unsanitary. Hand stripping should be done as rapidly as possible, otherwise the cow becomes a 'stripper' and only lets its milk down slowly. Cows milked by machine should be machine, not hand stripped.

Some cows are hard milkers even if they are properly managed. This is usually due to the fact that their teats have small orifices. As this condition may be inherited it is better not to breed from these animals.

In unimproved breeds the reflex to let down milk is initiated by the calf suckling and by that alone. It is a character which has been eliminated in the improved breeds and in all but a few good dairy cows of any breed. A cow which has been sympathetically handled by one attendant before calving will generally respond to hand milking by that attendant after calving.

Any cow, however, which has been accustomed to suckle her calf or to have it with her will not let down her milk in its absence. Cows are usually milked twice a day at approximately 12-hour intervals. High-producing cows may with advantage be milked more frequently.

An increase of more than 10 per cent may be expected by milking three times daily. Experiments in Australia and New Zealand have shown that contrary to previous ideas there is no change in the average hourly excretion of milk at milking intervals of between 8 and 16 hours. The increased production obtained by milking three times a day is not obtained by a higher excretion rate per hour but is due to other causes.

Hand versus machine milking

Although only a small proportion of all the cows in the tropics are machine milked, it can be assumed that the numbers will increase.

It is generally accepted that machine milking is not quite as good as the best hand milking but that it is better than average hand milking. Cows like machine milking as well or better than hand milking. Machine milking can be introduced if labour is scarce and expensive, if available labour dislikes the toil of hand milking, and if enough cows are milked to justify the installation of a machine.

It is usually considered, in the temperate zone, that the milking of twelve cows justifies the installation of a small machine. Even if a farmer has a herd large enough to justify machine milking, he is not likely at present to machine milk in most tropical countries, as this may be more expensive than hand milking, due to the low cost of labour.

Anker-Ladefoged has shown that in the U.K. the production of one gallon (4.5 litres) of milk will pay for two-thirds of an hour of labour, while in Ceylon it will pay for ten hours.

There are two main types of milking machine, bucket and line or releaser machines. In the temperate zone bucket-type machines have been most popular in those countries where the cows are managed indoors for part of the year such as Denmark, England and the United States, while releaser plants have been almost universal in countries where the cows are managed outdoors throughout the year such as New Zealand and Australia.

The present tendency is for releaser-type machines to increase in popularity in all countries as they require less labour. Although almost all the machines installed in tropical countries in the past have been bucket machines a farmer installing a machine today would be well advised to install a releaser machine.

The essentials of a milking machine are a vacuum pump to create the vacuum, a vacuum tank, a vacuum line to the individual sets of teat cups, a method of collecting the milk and a pulsator. Vacuum pumps can be of the piston or rotary type and can be driven by a small petrol engine or an electric motor.

It is essential to see that there are no leaks in the vacuum system. The normal teat cup is made of metal or plastic and has a rubber inflation or lining. Two types of inflation are in general use. The ordinary rubber liner which collapses and closes under the teat and the moulded inflation which collapses around the teat, and so never obstructs the flow of milk.

The pulsator is a device that produces an intermittent vacuum, causing an alternate compression and release of the cows teats through the action of the rubber liner or inflation in the teat cup. The function of the pulsator is first to apply a stimulus to the teat to ensure that the milk is let down, and second to massage the teat in order to maintain circulation of the blood.

Milking without pulsation causes the teats to swell, makes the cow uncomfortable, and thus upsets the let down of the milk. There are several different types of pulsator. All the milk that has been let down can be removed by the proper manipulation of the teat cups and by massaging the udder. Although the odd cow may have to be hand stripped, every effort should be made to strip all cows by machine.

As milking progresses, the lower part of the udder becomes soft and flabby due to the release of pressure caused by the let down of the milk. These soft tissues tend to be gradually drawn into the teat cup and this slows down milking and finally completely blocks the passage of milk from the gland cistern to the teat cistern.

If the teat cups are drawn down with one hand while the udder is massaged with the other, all the milk that has been let down can be withdrawn. Machine stripping is desirable because it saves labour, stops the cows from becoming 'strippers' and reduces the danger of milk contamination.

The quality of milk obtained by machine milking depends upon the care taken in cleaning and operating the machine. If the machine is properly cleaned, then the milk will certainly be of as high a quality as milk produced by the best hand milking. Machines and other dairy equipment may be sterilized by steam or by chemicals. Milking machinery gets a deposit of 'milkstone' on the metal and rubber, and fat is absorbed into the rubber. 'Milkstone' is a deposit consisting of milk solids together with mineral constituents from the water used in washing the machine.

It increases the tendency of the rubber to crack, reduces the efficiency of the rubber rings, and harbours bacteria. Fat absorption causes rubber to lose many of its important qualities when exposed to light. As a consequence milking machinery must be particularly well cleaned. Many different routine measures are recommended by different authorities, but any system to be successful must be very thorough.

New types of line milking machines are fitted with automatic washing equipment and these are recommended for use in the tropics where labour is often unskilled.

It is sometimes claimed that machine milking increases the incidence of mastitis in the herd. Misuse of a machine can certainly cause severe damage to the tissues of the udder.

The following milking rules are applicable to' both hand and machine milking:

(1) Avoid excitement for the cows either before or during milking. The cow should be relaxed and enjoy being milked.

(2) Stimulate let down of milk one or two minutes before milking begins. The most effective stimuli are the massage of the teats with a warm cloth in order to clean the udder, and the use of the teat cup. Milking should not be commenced until the milk has been let down. This can be seen by the distension of the teats and the lower part of the udder and should take place about two minutes after the initial stimulus.

(3) In hand milking use dry hands and the full-hand grasp of the teat. In machine milking see that the machine is thoroughly clean and in good order, that there are no leaks in the vacuum, that the pulsator is set correctly and that the inflations are not broken or distorted.

(4) With machine milking begin stripping as soon as the teat cups start crawling up the teats, or if a Ruakura sight glass is fitted, as soon as the sight glass clears. In hand milking, milk as rapidly as possible and avoid prolonged stripping.

(5) Remove the milking machine as soon as the milk stops flowing. There is no difficulty in doing this if a Ruakura sight glass is used.

It is usual to restrict milking gradually in order to dry off a cow. The pressure developed in the udder soon stops milk secretion. Cows should be dried off at 305 days even if they are producing well, so long as they are carrying a calf and are going to calve down within a 12-month period. Cows should only be dried off before 305 days if they are due to calve or if they produce so little milk that it becomes uneconomic to milk them.

Sometimes a cow's teat leaks milk before milking. This may be due to two conditions. The muscles in the orifice may be weak, or there may be a natural fistula or wound in the teat cistern. A wound may be sealed by cauterizing it or by surgical methods when the cow is dry. If teats are sore they should be bathed in a warm salt solution and vaseline applied.

Feeding

Cows managed outdoors should be rotationally grazed and not set-stocked. The length of the grazing cycle will depend upon local conditions but it is best to move the cows daily. The most efficient grazing system is known as strip grazing or close folding. Under this system the cows are given exactly the area that they need for one grazing, and this is done by dividing up fields by the use of some form of convenient fence.

The portable electric fence is very suitable for this purpose but very few peasant farmers in the tropics could afford to purchase this equipment. A small farmer can tether his animals and move them twice daily according to a set pattern. Experiments in Scotland have shown that the production per acre is markedly higher using a strip grazing rather than a rotational grazing system.

There is no reason to believe that the same results would not be obtained in the tropics, particularly in the humid tropics where grass growth continues throughout the year. The disadvantage of using this system is that management has to be very skilled.

Cows should always be placed on the best grazing at night as experimental work has shown that temperate type stock graze mainly at night and rest for long periods during the day. It is likely that high-producing tropical type cows will behave in the same manner, though such cows in Ceylon when grazed under coconut palms feed mainly during the daytime. Grazing paddocks should always be provided with shade, as should yards.

Little is known as to the value of most tropical pasture for milk production. It can be assumed that in the humid tropics good pasture will provide for the maintenance requirements of the cows and for the production of one gallon (4.5 litres) of milk.

Cows that are producing more than one gallon (4.5 litres) of milk should be fed concentrates according to the scale in Table 10. In the more arid tropics the pasture will usually only provide for maintenance and for some part of the milk production during the rainy season. At other times of the year dairy cows will have to be fed a supplement of cut fodder, hay or silage in the yard or on the pasture.

Table 7.5 : Scale of concentrate feeding to cows kept on good grazing in the humid tropics.

Daily milk production lb. (kg.)	*Amount concentrates fed lb. (kg.) per day*
0-10 (0-4.5)	0
10-15 (4.5-6.8)	2 (0.9)
15-20 (6.8-9.1)	4 (1.8)
20-25 (9.1-11.3)	6 (2.7)
25-30 (11.3-13.6)	8 (3.6)
30-35 (13.6-15.9)	10 (4.5)
35-40 (15.9-18.1)	12 (5.4)

The nutrient requirements for maintenance of cows of different weights are given in table elsewhere in this chapter. Morrison recommends that dairy cows require from 1.5 pounds (0.68 kg.) of dry matter for each 100 pounds (45.4 kg.) body weight when they are dry, to 2.1-2.5 pounds (0.95-1.1 kg.) when they are producing one pound (0.45 kg.) of butterfat per day, to three pounds (1.4 kg.) when they are producing two pounds (0.91 kg.) of butterfat per day.

Tropical type milkingcows usually produce milk with a higher total solids content than that of temperate type cows, and approximately four pounds (1.8 kg.) of a concentrate mixture containing at least 3.10 to 3.25 (1.4-1.47 kg.) of S.E. and 0.77 to 0.8 pound (0.35 to 0.36 kg.) of P.E. are required for the production of one gallon of milk.

The variety of concentrates available for the feeding of dairy cattle in the tropics is very large so that it is impossible in a general treatise to recommend rations that are available in all areas. A ration that has been found to be very suitable in one area is the following:

Groundnut meal	30 parts
Coconut meal	30 parts
Maize meal	20 parts
Rice bran	10 parts
Molasses	10 parts

If it is possible each cow should be fed its ration separately, the amount varying with its production. If cows are managed outdoors, concentrates are best fed at milking time, otherwise they should be fed twice a day, early in the morning and as late in the afternoon as is convenient. If concentrates are fed at the bail, automatic feeding devices can be installed.

Table 7.6: Nutrient requirements for the maintenance of dairy cows.

Live-weight lb. (kg.)	*Digestible crude protein lb. (kg.)*		*Total digestible nutrients lb. (kg.)*	
	Minimum	*Good*	*Minimum*	*Good*
700(317-5)	0-440(0-200)	0-476(0-216)	5.13 (2.33)	5-81(2-64)
800(362-9)	0-494(0-224)	0-536(0-243)	5-77(2-62)	6.53 (3.08)
900(408.2)	0-547(0-248)	0-593(0-269)	6-38(2-79)	7.23 (3.28)
1,000(453-6)	0-600(0-272)	0-650(0-295)	7-00(3-18)	7.93 (3.60)

Cows should always have access to water in the paddocks and in the yards. The consumption of water by dairy cows in the tropics is at least twice what it is in the temperate zone (Payne and Hancock, 1957) and the necessary arrangements should be made to see that cows are able to obtain sufficient water.

A complete mineral mix should be added to the rations or fed separately. Details of a suitable mixture is given in other chapter of this book.

The Bull

The selection of a bull of the right type is very important as a sire is 'half the herd', that is, half the inherited characteristics of all the calves are obtained from him. Bulls may be selected by type, pedigree or by sib or progeny performance. The relative merits of each type of selection have been discussed in other chapter of this book.

Temperate type bulls have often been imported into the tropics in the past to be used for grading-up purposes. If this is to be done, and there are some parts of the tropical world where it is still a practical method of improving milk production, then the bulls' ability to tolerate heat should be tested before they are imported.

At least this will ensure that they have a better chance of surviving in the tropical environment into which they are imported. This has been attempted in at least one tropical country with favourable results. The method used has been to select at least six bulls of equal merit in the temperate zone and to test them: the bull that demonstrates the most tolerance to hot conditions being the one selected for importation.

Another precaution that should be taken is to select bulls that shed their winter coat readily in the spring. It is quite certain that a bull that retains his winter coat in the summer in the temperate zone will be quite useless if imported into the tropics.

Management and feeding

The general feeding and management of young bulls should be similar to that of heifers. Every effort should be made to feed them particularly well so that they can be brought into service at as early an age as possible. When fully grown they should be kept in a vigorous physical condition and should not be allowed to get too fat. On peasant holdings the bull may be worked lightly, but it is not usual to do this on larger holdings.

When mature the bull should be fed approximately the same as a dry cow but should receive daily an extra two to four pounds (0.91 to 1.8 kg.) of the milking cow's concentrates. If he is not run with the dairy herd, which is advised, he should be kept in shady paddocks and provided with ample water.

Bulls managed in the tropics appear to be quieter and less bad tempered than temperate type bulls managed in the same way in the temperate zone. Bull paddocks should always be better fenced than the cow paddocks.

If the bull is not tractable, a ring should be inserted in his nose. This is done by punching a hole in the septum with a special pair of pliers or some other sharp instrument, and fitting a copper ring of suitable size. Dehorning will quieten any bull.

Daily exercise is most important for bulls. If they are kept indoors and do not have access to a paddock they may be led at a good walking pace for a mile or two in the cool of the evening.

Service management

Bulls vary in the length of their active life from 5 to 16 years. The age of a bull has no influence on the quality of his offspring. In the temperate zone young bulls can be first used at 10 to 12 months of age but in the tropics temperate type bulls are usually not well enough grown for service before they are 18 months old.

Tropical type bulls are not usually ready for first service before they are two or more years old and do not usually reach their maximum of breeding power until they are three to six years old. It is possible that the breeding power of bulls in the tropics does not decline at so early an age as it does in the temperate zone.

Young bulls can be used once or perhaps twice a week, but older bulls may be used up to five times a week. An adult bull of certain breeds can serve up to 100 cows in a year but usually a bull is allowed 50 to 60. Very large cows should be placed in a shallow pit for service by very young bulls and small cows and heifers should be served in a service crate by very heavy old bulls.

There is some evidence from sub-tropical areas that the fertility of temperate type dairy bulls is lowered during very hot weather. It is not yet known whether tropical type bulls behave in the same way, but as a precaution every effort should be made to ameliorate the climatic conditions for all bulls in the tropics by good management, using the techniques suggested in a previous section for the proper management of milking-cows.

BREEDING

The details of the subject are fully discussed by Mahadevan (1958). It has been stated in a previous section that there is no single breeding policy that is applicable in all tropical countries. All over the tropical world the value of local breeding policies should be reassessed with special reference to the new physiological knowledge on the reaction of dairy cattle to climatic stress.

There is in fact a choice of three possible major breeding policies, either or all of which might be useful in a tropical country attempting to breed more highly productive dairy cattle. These are : selection for productive ability in well-adapted but not very productive indigenous tropical breeds; selection for adaptation in ill-adapted but highly productive temperate breeds; and the crossing of suitable tropical and temperate breeds and subsequent selection and interbreeding of crossbreds that are characterized by high production and adaptability.

The first policy is most suitable for countries in Asia and Africa, where there are a large number of indigenous tropical breeds, and where standards of management and feeding are generally poor. It is obviously desirable that everything should be done to improve the productivity of these indigenous breeds and to preserve even tropical breeds that at present have no apparent economic utility.

It must not be expected, even with the most suitable breeding techniques, that productivity in the indigenous tropical breeds will rapidly improve. In Ceylon Mahadevan (1951) has estimated that it will be 200 years before the productivity of the indigenous Sinhala breed could be raised by selective breeding to the same level as that of some of the exotic Jersey cattle imported into the island.

The second policy is most suitable for the tropical areas of Australia and the Americas, where there are no indigenous breeds and where there are already comparatively large numbers of temperate type cattle. It is probable that there has already been some natural selection pressure on the temperate type stock in these areas.

The Criollo, the typical cow in Central and South America, is a descendant from the original temperate type stock imported from Spain. It has been considered a degenerate animal but in many respects it has evolved towards a type suited to the tropics as it is now resistant to tick-borne disease and to the local warble fly and is apparently quite well adapted to climatic stress.

There is indeed considerable evidence to suggest that the variation in adaptability to a climatic environment of individual animals within temperate breeds, is as great as the variation between tropical and temperate breeds. The practical proof of this thesis is that there are today in Australia, the Pacific Islands, the Americas and the Caribbean a number of herds of temperate type cattle that although not yet fully acclimatized, produce as well as, or better than, the very best herds of indigenous tropical cattle in Asia or Africa. The most common temperate type breeds found in the tropics today are the Holstein, Jersey, and Brown Swiss.

The crossbreeding of tropical and temperate type cattle and subsequent stabilization of a crossbred type retaining both productivity and adaptability, is superficially the most attractive of the three policies. Attempts to grade up tropical with temperate type stock should not be confused with this policy.

Many attempts have been made to grade up tropical stock with temperate stock but few serious attempts have been made to crossbreed and then breed *inter se*. Crossbreeding experiments in the tropics have been reviewed by Maule who has shown that much of the work has been inconclusive. An attempt has been made in the United States to use this method to produce a dairy animal suitable for the southern regions of the United States. Although theoretically this is an attractive policy it is a difficult one to execute in practice.

General Considerations for the Farmer

The major problems associated with breeding productive dairy cattle for both the humid and arid tropics are not usually the concern of the average dairy farmer. His problem is to use to best advantage the dairy animals available in his locality, until such time as his government or a large scale breeder introduces superior stock.

Whatever type of stock he rears, tropical, temperate or crossbred, there are certain breeding management practices that he can apply that will ensure that he is making the best possible use of them. First he must manage and feed his cattle as well as possible so that they have a chance to express their productive ability.

This means good feeding and management from birth and not just when milking. Then he must record the productive ability and other characteristics of his cows and from the records select the highest-producing and most desirable cows for his future breeding programme and cull the remainder.

Finally he must select a suitable sire. This was very difficult for the small farmer before the advent of artificial insemination, but now in those areas where there is an artificial-insemination service he should be able to procure semen from superior sires. If there is no artificial-insemination service he should select his future sires from the progeny of the most productive cows.

The future female breeding stock should be selected for high and persistent yield of milk and butterfat, regular calving, longevity, and a good dairy temperament. It has been demonstrated that it is best to select for all these characters at the same time, rather than one at a time.

The yield of milk and butterfat cannot be estimated by looking at the cow but only by recording production. The milk may be weighed daily, weekly, monthly or only three or four times during the lactation. A monthly record is sufficient to make an approximate estimate of production and a bi-monthly record will be of some use.

Regular calving at 12-month intervals is an index of fertility. It is important that the cows calve at regular intervals as they are likely to be more productive in the first three months of a new lactation than they will be if they milk for an additional three months at the end of a tenmonth lactation. Longevity is important because, other things being equal; the more lactations a dairy cow completes the more profitable she will be for the farmer. In the temperate zone the cost of raising a heifer from birth to first lactation is approximately the same as the value of the milk produced during the first two lactations.

As heifers grow more slowly in the tropics and on the average produce less when they are mature, the cost of raising a heifer in the tropics may be equal to the value of the milk produced during the first three lactations. This means that a cow that lives to complete six lactations is three times as profitable to the farmer as a cow that only lives to complete four lactations.

Dairy temperament is important as a vicious cow or a kicker will waste time at milking and upset the routine for the other cows. There is some evidence that dairy temperament is highly inherited.

If the productivity of cows cannot be ascertained by looking at them, it is quite certain that this method is useless for the selection of dairy bulls. They must be chosen after some form of test and details of these tests have been discussed in a previous section.

If yields are to be recorded, the cows should be marked by one of the methods detailed in Appendix II. The simplest and cheapest methods are to ear tattoo them as young calves and to brand or ear nick them when they are older.

It is easy to weigh the milk if the farmer is milking by hand or using a bucket milking machine, but it is more difficult when he is using a releaser or line machine. When a releaser milking machine is used it is usual to milk into test buckets at definite intervals. Various milk flow recorders have been invented but unfortunately none of them is very reliable.

There is one type of releaser recorder machine where the milk passes into glass jars suspended from a balance. This type of machine is accurate but very expensive. In official milk recording schemes milk is weighed at three-week or monthly intervals. The percentage of butterfat in the milk is usually determined at the same time.

There are two major methods used in determining the percentage of butterfat in the milk. These are the Babcock method used in the Americas and Australia and the Gerber method used in most other countries. Details of these methods are given in any textbook on dairying.

Different countries organize herd testing in different ways. In some it is organized by the Government, in others by the breed societies or by co-operative farmer organizations. Whatever the form of organization the aim at the local level is to present the farmer with knowledge of productive ability of his cows so that he can cull the low and breed from the high producers, and the aim at the national level is a systematic cataloguing of the country's superior stock from which can be selected sires to be used in artificial insemination or other improvement schemes. The subject is discussed in other chapter of this book.

Special Considerations for the Breeder in the Tropics

As animal-breeding techniques have been developed in the temperate zone, it has not been appreciated until recently that there are some special considerations that have to be taken into account by an animal-breeder in the tropics.

A tropical climatic environment imposes special problems. Dairy cattle should be selected for heat tolerance, resistance to disease, parasites and photosensitivity as well as for high production. If it seems unlikely that nutritional and management conditions can be improved, then the cattle should also be selected for tolerance of poor nutritional conditions and for hardiness.

These special conditions imposed by a tropical climatic environment dictate a physiological approach to breeding problems. The first requirement is the identification of characteristics desirable for high heat tolerance.

These may be: body and skin temperatures; respiration rate; body size and shape and distribution of skin appendages; size, density and functioning of sweat glands; length, colour and density of hair; colour of skin; standard metabolic rate; digestive efficiency, and other considerations. These problems are briefly discussed in other chapter of this book..

DAIRY FARMS AND EQUIPMENT

If a new dairy farm is being laid out the buildings should be sited as far as possible in the centre of the holding, so that if the cows are managed outdoors they will only have to walk a minimum distance to and from the fields and if they are managed indoors fodder cut in the fields will only have to be carried a short distance.

The grazing area should be subdivided with as many fences as is economic. In the humid tropics the field size should be no greater than two to three acres (0.8-1.2 hectares), though it may have to be much larger in the arid areas. A minimum of fourteen separate paddocks is required for rotational grazing in the humid tropics and rather more in the arid tropics. Lanes between fields should be narrow in order to facilitate droving of the stock. If the farm is mechanized, gateways must be large enough to admit machinery.

A dairy farm needs to be well drained but ditches should be fenced on both sides otherwise dairy stock will soon tread-in the sides. Water should be provided in all fields if this is practicable and economic. If it cannot be provided in the fields it should be provided at the milking shed so that the cows can drink as soon as they are brought out of the fields. Shade trees, of fodder varieties when practicable, should be planted around all the fields and along the droves.

Buildings

The type and number of buildings required will depend upon whether the dairy farmer manages his stock indoors or outdoors. If he manages them indoors he will require many more buildings, and if his farm is medium or large in size he will require a great deal more machinery.

Dairy stock managed indoors may be fed and milked in a dairy barn or they may be managed in loafing yards and milked in a separate shed. Barns or yards will also be required for bulls and heifers.

If the stock are kept outdoors only a milk shed for the cows is required and elaborate barns can be dispensed with. Milking sheds should have associated with them a small milk room, calf sheds, feed store and implement sheds.

The detailed design of a *milking barn* varies from country to country, but the general characteristics should be as follows. It should be sited with easy access to a road, adjacent to a water supply and convenient for movement of cattle, milk, fodder and manure. It should be built on as exposed a site as possible so that the maximum advantage can be taken of any breeze, and large shade trees should be planted round it.

Dairy barns may be of two types, single or double range. The double-range barn is more economical for herds of more than 16-20 cows. In double-range barns standings can be designed so that the cattle face inwards on to a central feeding passage, or outwards with feeding passages on either side of the building.

It is generally considered that it is more convenient for the operator and healthier for the cows to build the standings so that the cattle face outwards. A typical plan of a double-range dairy barn for medium size cows is shown in Figure elsewhere in this chapter.

The length of the standing for medium size animals should be five feet (1.53 metres), while for small ones such as the Red Sindhi it should be four feet six inches (1.37 metres). The width of a double standing

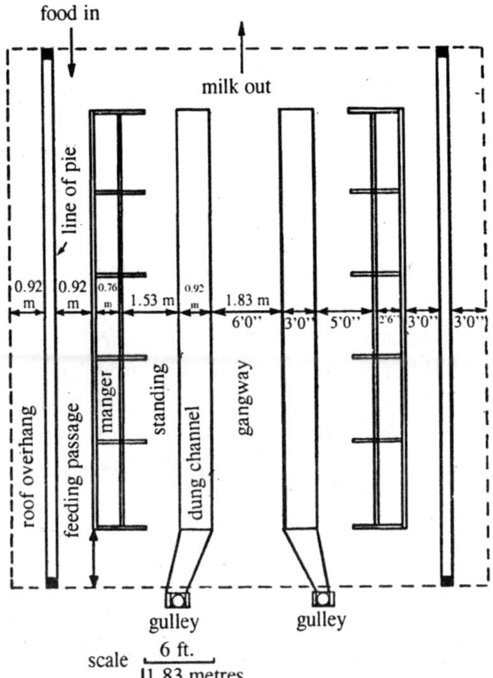

Figure 7.2: Plan of typical double-range dairy barn for medium size cows.

should be six feet six inches (1.98 metres) for small animals, and that of a single standing, three feet six inches (1.02 metres). The width of the manger depends upon the type of tie used; the maximum width should be two feet six inches (0.76 metres).

Feeding passages should be three to four feet wide (0.91-1.22 metres) and dung channels should be three feet (0.91 metres) wide if they are to be efficient. This is wider than is usually recommended. A dairy barn in many tropical countries does not need to be constructed with walls.

The roof should be high enough to allow for the use of all the equipment and it is best constructed of a good insulating material such as asbestos or thatch.

If corrugated iron is used it should be painted white on the outside and black underneath. There must be a wide overhang on the roof in order to prevent rain blowing into the barn.

The floor should be constructed of concrete which should be left with a rough finish in order to prevent the cows from slipping. The standings should be constructed of tubular steel and the mangers of glazed pipe. Water should be available to the cows in every standing, preferably from automatic watering devices.

In the temperate zone 30 gallons (136.4 litres) of water per cow per day are considered necessary for drinking, cooling milk and cleaning; more is required in the tropics.

Loafing yards do not need to be elaborate structures. They can be concrete, gravel or dirt yards with provision for shade, feeding, drinking and the removal of manure. In the humid tropics, concrete or gravel yards are essential but elsewhere a floor of rammed earth will often suffice.

Shade should be adequate and 40 to 60 square feet (3.7-5.6 square metres) are needed per cow. Water should be provided preferably from automatic devices, but if not, from a trough under shade. Feeding and manure clearing can easily be mechanized in yards of this type. When cut fodder or silage is fed, each cow requires about two feet six inches (0.76 metre) of trough space.

The yard fences are best constructed of concrete posts and steel cable, but any type of good fence will suffice. There should be a number of separate yards for the segregation of milking cows, dry cows, heifers and bulls. The yard may be laid out rectangularly or in a semi-circle. There are advantages in both method,

The *milking shed or bail* is used by farmers who'manage their cows outdoors or indoors in yards. It is sometimes also called *a milking parlour.* It is a shed built specifically for milking, and a dairy room, feed store and assembly and dispersal yards are usually associated with it. There may also be a separate section used for washing the cows before milking.

A number of variations in design are possible, and all types can be used for hand, and bucket or releaser machine milking. The most common types are as follows: (1) milking-barn type where the cows are tied in stanchions, (2) single-level herringbone type, (3) single-level, abreast, walk-through type, (4) two-level, abreast, walk-through type, (5) two-level, tandem, walk-through type.

The milking barn parlour is essentially the same as a normal milking barn except that part of the barn is reserved for milking purposes only. The cows are milked in batches. In the herring-bone type the cows are also milked in a batch, but they are placed side by side at an angle in two rows with the heads of the cows in each of the two rows towards each other. This is a more efficient way of 'batch' milking than the ordinary barn but is little used. The single-level, abreast, walk-through type is the most common and when well planned is probably the most labour saving.

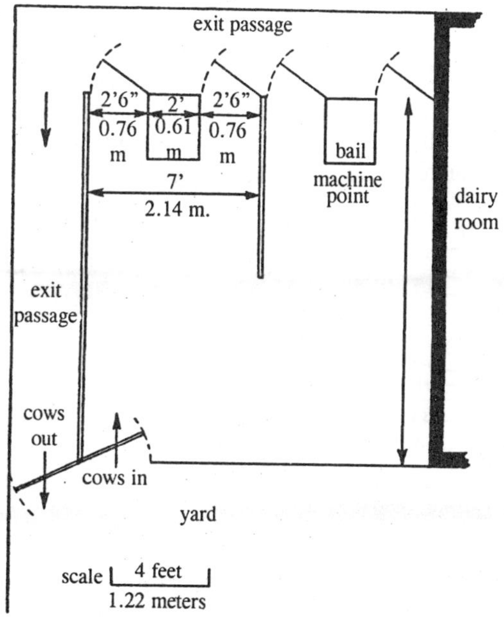

Figure 7.3: Plan of a four-stall, single-level, abreast type of milking parlour.

In New Zealand where this type of layout is usual it is not uncommon to find one man milking more than 60 cows. A ground plan of a four-stall, singlelevel, abreast type of milking parlour suitable for the tropics is shown in Figure elsewhere in this chapter.

Where a line milking machine is used the stalls should be two feet six inches (0.76 metre) and the bail two feet wide (0.61 metre), whereas when a bucket milking machine is used or hand milking is practised the stalls should be up to three feet three inches (1 metre) wide.

It is usual to have one milking machine unit at each bail, so that one cow can be milked while the cow on the other side of the bail is being washed. Hot and cold water can be on tap at the bail for washing purposes, and concentrates can be fed at the bail by hand or by using automatic feeding devices.

A twounit system is the economic minimum. It is considered that for up to fifteen cows a bucket machine is more economic to use in this type of milking parlour while above sixteen cows a line type of machine is more economic. The floor should slope away from the bails either in front or behind the cows. The two-level, abreast, walkthrough type is a variant of the previous type, where the milker works at a lower level than the cows stand.

It is not very usual and does not possess any outstanding advantages. In the two-level, tandem, walk-through type the milker works in a pit while the cows are milked in stalls set in tandem and arranged head to tail. The stalls may be in one row, arranged around a square or in banks, the milker working in the centre.

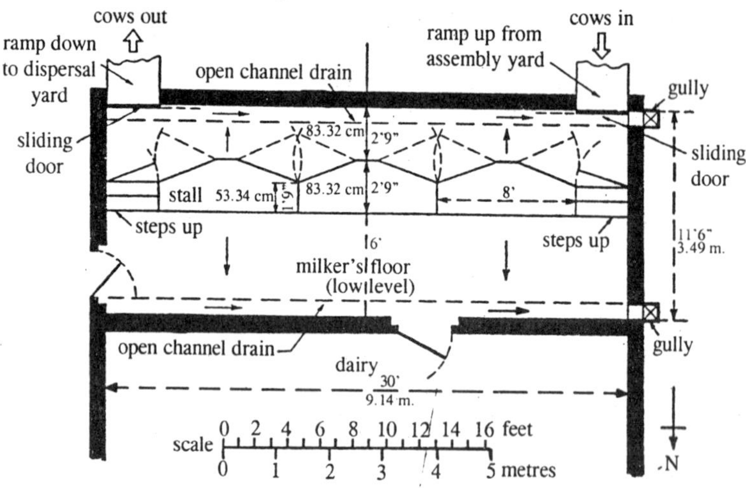

Figure 7.4: Plan of typical tandem-type milking parlour.

The difference in level between the cow and the milker varies between one foot nine inches (0.53 metre) and two feet six inches (0.76 metre). The length of each milking stall is eight feet (2.44 metres). Two gates are necessary. A ground plan of a typical tandem-type milking parlour is shown in Figure elsewhere in this chapter.

This type of milking parlour is more expensive to construct than the others and it requires associated washing stalls as washing in the milking stall decreases the through-put of cows. A separate milking unit is required for each stall and it is more difficult to feed concentrates by hand. It is really only suitable for very large dairy farms.

The milking parlour in the tropics only requires a roof. Provision should be made for shade and water for the cows while they are held in the yards and sprinkler systems can be installed to add to the comfort of the stock.

A small *dairy room* is needed whether the cows are hand or machine milked. If the cows are hand milked or milked with a bucket machine all that is required is a small room adjacent to the milking barn or parlour, that can be easily cleaned and is screened to keep out insects. It should be sited on the side from which the prevailing wind blows and on the other side from the feed store.

The milk will be strained in this room and the milk utensils washed and cleaned, so that provision for hot and cold water is desirable. Cooling equipment is also desirable and if it is economically feasible refrigeration equipment.

If the cows are milked with a line plant the releaser part of the equipment should also be in the dairy room which must therefore be sited at the end of the line.

The size of the feed stores will vary according to that of the farm and the types of feeds used. All feed stores should be constructed so that rodents are excluded and so that they can be easily fumigated to destroy weevils.

The essentials of a good calf shed have already been described in a previous section. Dairy cattle yards do not have to be so stoutly constructed as beef cattle yards. Concrete or steel upright posts are best and the most suitable materials for the horizontal bars are tubular steel or steel cable.

In collection or dispersal yards, at least 25 square feet (2.3 square metres) should be allowed per each polled cow, and 40 square feet (3.7 square metres) per horned cow.

Milking Equipment

When hand milking the dairy farmer requires only a minimum amount of equipment. Seamless milking pails, washing buckets, a teat cup and a stool together with straining, washing and sterilizing equipment in the dairy room. If the farmer is recording, a simple milk weighing scale is also required.

The majority of peasant dairy farmers in the tropics do not possess as much equipment as this and the aim should be to provide them with these minimum requirements and teach them how to use and clean the equipment properly.

A great deal more equipment is required if milking machines are used, but machines differ in their complexity. The essentials of a milking machine have already been described in a previous section. If a medium or large-scale dairy farmer can afford and considers that it is economic to install milking machinery, then undoubtedly the most suitable and efficient type is the 'releaser', constructed in stainless steel and incorporating automatic cleaning equipment. The dairy farmer in the tropics requires a machine that is simple in construction and operation, needing the minimum of servicing and easily cleaned by relatively unskilled labour.

A machine that fills these specifications has been described by Phillips (1956) and details of the machine layout are shown in Figure 18. This machine has a 'master' vacuum-operated pulsator system that controls relay or 'slave' pulsators in the milking shed. No lubrication or separate electrical fittings are required.

The vacuum pump has a high pumping capacity and is driven directly by an electric motor, thus eliminating belts and pulleys. It may of course be driven by a petrol engine. The most notable feature of the machine is a semi-automatic washing system that circulates washing solutions through the plant when milking is completed. On the bajj, directly coupled into the milk line, the Ruakura milk flow indicator is used.

This is a simple device that indicates to the operator when milking is ended and stripping should begin. It has been found to be particularly suitable for use in the tropics where milking labour may be relatively unskilled.

Equipment for the Dairy Room

Straining

All milk should be strained immediately after the milking process is completed. This can be done by passing it through a variety of materials. Special strainer pads are the most efficient and hygienic as they can be discarded after use. Any type of suitable cloth can be used but care must be taken to see that it is changed frequently in use and thoroughly washed and sterilized after use.

Cooling and refrigeration

As soon as the cow has been milked the bacteria in the milk start multiplying. Immediate cooling of milk or cream reduces bacterial multiplication very drastically. In Table 12 details of the relation between the temperature and the bacterial count of milk 12 hours after milking is shown.

Some of the bacteria produce chemical changes in the composition of the milk, the most common one being the formation of lactic acid from lactose or milk sugar. This causes the so-called souring of milk. As air temperatures are high in the tropics, milk easily and quickly

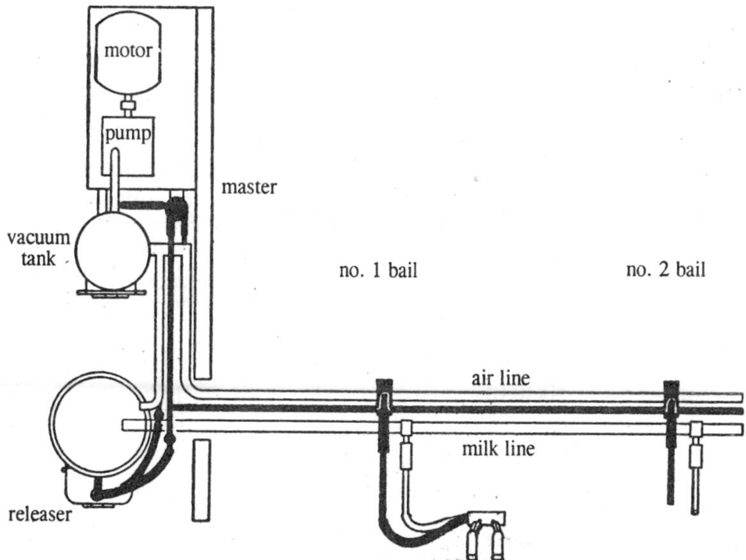

A sketch showing the layout of the plant witht he master and relay pulsator system marked in black.

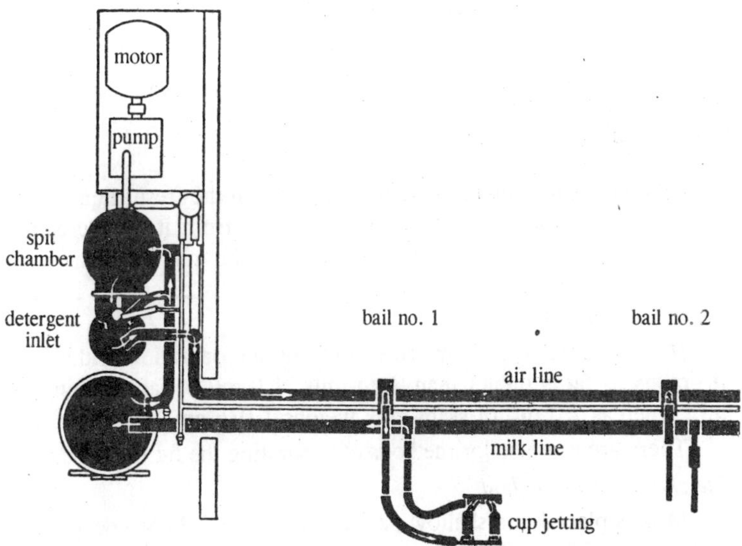

The path (in black) taken by the wasing solution during cycling. The cleaning solutions are drawn in near the vacuum tank. After circulation they are once more released into the bucket for further circulation.

Figure 7.5: Plan of a line milking machine suitable for use in the tropics.

sours unless it is cooled or boiled. Boiling kills the bacteria, but it must be carried out before the concentration of lactic acid in the milk has risen too high. Thus if fresh milk is going to be distributed in the tropics, it must be cooled immediately after leaving the cow's udder.

Table 7.7 : The relationship between temperature and bacterial count in milk 12 hours after milking.

Temperature °F. (°C.)	*No. bacteria per ml.* Thousands
40(4-4)	4
45(7-2)	9
50(10-0)	18
55(12-8)	38
60(15-6)	453
70(21-1)	8,800
80(26-7)	55,300

This is a very difficult problem in the tropics where the farmer usually has a small number of cows and cannot afford refrigeration. It is discussed in other chapter of this book.

A solar water cooler may be used. This is a device that takes advantage of the cooling effect of the evaporation of water and is most effective in hot, dry climates.

Separation

When milk for any reason cannot be sold in the fresh, liquid state, it is generally desirable to remove the cream from it. The cream is used for the production of ghee or butter, and the milk for a variety of purposes. The making of these milk products is discussed in other chapter of this book.

Milk fat rises to the surface of milk by the operation of the force of gravity as fat is lighter than skim-milk, 1.0 ml. of fat weighing 0.9 gm., whereas 1.0 ml. of skim-milk weighs 1.036 gm.

There are three major methods of separating the fat from the milk:

The shallow pan method

Milk is placed in a shallow pan four inches (10.2 cm.) deep for 24 to 36 hours. The milk turns sour and the cream rests on top and is removed with a skimmer. The skimmilk left in the pan usually contains one-half to one per cent butterfat.

The deep pan method

Milk is placed in a pan 20 inches (50.8 cm.) deep and, after standing some hours, the cream is removed by a dipper. Sometimes the milk is diluted with water before being placed in the pan, as it is claimed that dilution lowers the resistançe of the fluid to the rising fat globules.

This is a more efficient method than the shallow pan one, but if it is to work really satisfactorily the deep pan should be allowed to stand in a bath of cold water, which is not usually a practical proposition in warm climates.

Mechanical separation

This is carried out by applying centrifugal force to the milk. When milk is spun rapidly in a container the force of gravity is increased by about 1,000 times so that the difference of 0.136 in the specific gravity of butterfat and skim-milk at rest, becomes a difference of 136, and the fat separates out almost instantly from the skim-milk towards the centre of the container in which it is being spun.

Dispersing the milk in the container into thin sheets facilitates separation and all modern milk separators incorporate a series of inverted cup-like discs inside the bowl, each kept slightly apart.

Mechanical separators are made in almost any size, farm separators usually being designed to deal with 10 to 100 gallons (45.5 to 454.6 itres) of milk per hour.

BEEF CATTLE

Beef is produced in the tropics almost entirely from cattle on free range using the grass and herbage growing on land which for one reason or another has, so far, been found to be unfit for any other form of agricultural production. All that beef cattle are permitted to use from cultivated land are the residues of crops, when these are not required for other stock.

The question arises, are cattle the best converters for this purpose ? In tropical East Africa it has been shown (Ledger, 1961) that indigenous wild ungulates have specific advantages over domestic species in this respect as judged by the relative amount of lean meat produced.

The carcase of *Bos indicus* has an average of 32 per cent while that of Thomson's gazelle has 46 per cent. There is practically no intensive beef production such as is practised in the temperate regions of the world where early maturing cattle are grown on first-class pasture or on specially cultivated crops, and stall-fed on material designed to exploit their ability to fatten and finish quickly.

The owners of tropical beef cattle are either those whose livelihood depends primarily on beef production or those to whom the production of meat is incidental to other considerations.

The former group is composed very largely of people of European descent and outlook, in Australia and the Americas; the latter group consists of pastoralists who, in the process of increasing the number of their stock find themselves short of cash or with the prospect of loss through overstocking in drought or famine, and the settled farmers whose cow has become barren or work-ox too old, or who, stimulated by a local demand, occasionally undertakes the purchase and fattening of an ox or two.

In considering the production of beef, therefore, one might be disposed to concentrate attention on the interests of those whose primary object it is, and to ignore the others, but actually the amount of beef produced as we have said, almost incidentally, is very great and provides the only beef, and often the only meat, for millions of people in tropical Africa and Asia.

The supply, however, is inadequate even for those local needs and, being far short of what it might be if other methods were used, it merits full consideration.

The real beef producer's aim is to sell each year the maximum weight of beef cattle which he can raise on his property. In the tropics he is not only a rearer and seller of store cattle, that is, those which are grown and suitable for fattening on richer feed, but he himself must produce the cattle finished and ready for slaughter.

He has neither the option of turning his stock over to the arable farmer with surplus grain to feed nor to stall-feed them himself. His property is often situ in territories where the rainfall is so light that cultivation, except in very restricted areas, is not feasible, and where therefore vegetation of any description is scant, and surface evaporation high.

If he is to make a living which will support him and his family at what he considers an adequate level, he must maintain sufficient head of stock, and as the area required to keep each animal is often as much as 20 to 30 acres (8.0 to 12.1 hectares) and sometimes as large as 100 acres (40.5 hectares) or more, the total area of his property is correspondingly large.

Under such circumstances roads and other means of communication are expensive and inadequate, and constitute still another reason why he must himself be breeder, rearer and fattener. Because of the dispersion of the stock, great distances have often to be covered before they can be concentrated in sufficient numbers to justify the organization of a marketing centre where facilities for dealing with them are adequate.

The roaming pastoralist lives under very similar, or even harder, climatic conditions but, not being bound to the land nor dependent upon permanent sources of water, nor being particularly interested in raising stock for sale, he moves with the rainfall and the pasture, and his cattle may, on the whole, be put to less stress than are those of the rancher.

He can take advantage of the crop residues on cultivated land he may contact and, in many territories, he can leisurely direct his movements towards any suitable market without much inconvenience to himself or strain on his cattle.

The producer whose main activity is the cultivation of land is not often in a position to breed or rear cattle but in favourable seasons can profitably fatten a store bullock or finish the growth of an immature beast on surplus straw and other by-products of his farm not required by his permanent stock.

What feed will be available for that purpose is seldom determined beforehand but is a matter of chance depending upon the fortunes of the year. He is the only class of producer who might be in a position to practise the intensive methods used in the temperate zones, but with few exceptions the market he has to supply at present is one exclusively for cheap meat, and his choice of feeding materials and methods of management are too severely restricted to allow him to cater for any other.

The first category of owners very often introduced temperate type cattle and later used improved temperate type beef breeds. These breeds have been satisfactory just in so far as the environment continued to resemble that in which they had been evolved.

As long as the food supply was adequate and fairly easily procured, animals of these breeds could out-do those of tropical ones on the return they gave for it, but in times of scarcity they lacked the power required to carry them through without undue loss of weight or life.

There are few beef cattle areas in the tropics where the food supply is not inadequate during at least four months in the year and only those cattle which can not only survive but maintain themselves in reasonable condition throughout that period give a worthwhile return.

In this one respect the tropical breeds excel but, when the food supply again improves, it is the cattle of specialized breeds, if they have survived, which most quickly pick up weight and continue their growth. What is required and continually sought for is the animal which combines the valuable qualities of both classes of breeds. In tropical Australia, British breeds have been found sufficiently satisfactory and Shorthorns, Herefords and Devons are still used in the greatest number, but crossbreeding with zebus is gaining in popularity.

The following data given by Mawson (1956) convey a fair idea of the relative values. They relate to a group of seven- to eleven-month-old Shorthorn and Shorthorn x Hereford steers compared to a similar group of three-eighths Brahman and five-eighths Shorthorn or Shorthorn x Hereford.

Over a period of two-and-a-half years, the first group gained on the average 0.66 pound (0.30 kg.) per day, while gain in the other groups was 0.76 pound (0.34 kg.). The dressed carcase of the zebu crossbreeds averaged 90 pounds (40.8 kg.) heavier than those of the other group and their dressing-out percentage was 51 against 49.

In Rhodesia, it was found that although the actual carcase weight of Hereford type cattle and Hereford x Africander was, on the average, 30 pounds (13.6 kg.) heavier than those of pure indigenous cattle produced under the same conditions, because of mortality and lower fertility, due to environmental stress, the indigenous cow produced on the average 287 pounds (130.2 kg.) of beef per annum, while the Hereford type produced only 242 pounds (109.8 kg.).

In tropical and sub-tropical America, to which European breeds were introduced first, the zebu has been much utilized, though some acclimatized strains of temperate type cattle have also been evolved. The American Brahman and the Indo-Brazilian are breeds of tropical type cattle that have been evolved from several different breeds of Indian cattle.

The Caracu and the Criollo are acclimatized breeds that have evolved by a process of natural selection from the original importations of Iberian cattle, while the Santa Gertrudis, the Beefmaster and the Jamaican Red are more or less stabilized crossbreds.

The Braford (American Brahman × Hereford), Brangus (American Brahman × Aberdeen Angus) and Charbray (American Brahman × Charollais) are as yet unstabilized crossbreds that nevertheless can be differentiated as separate types. The extent to which any of these new breeds or types are used depends upon local environmental conditions and on their availability in any particular area.

The African itinerant beef producer has evolved breeds peculiarly suitable to his circumstances, which are so different from those required by the improved beef breeds that one cannot well be compared to the other; but, in tropical Asia, where work and milk production are the main objects of cattle-breeding, the types nevertheless are well fitted to produce beef in quantity as far as circumstances allow.

In considering what constitutes a good beef beast, some facts concerning the nature of 'meat' and how it is produced should be borne in mind. 'Flesh' is essentially muscle fibres. These fibres differ in thickness or 'coarseness' according to the situation occupied by the muscle they compose and the work required of the muscle.

The growth of the muscles occurs mostly as a result of increase in size and thus in coarseness of each fibre, and so it is that the larger the muscle and the older the animal, the tougher and less tender is the meat.

The quality of the meat is associated with the fineness of the muscle fibres and the proper disposition of fat, and in these respects the young, early maturing animal excels.

The calf's body is composed of a high proportion of bone and offal in relation to flesh and fat, and as the calf grows, the fleshcarrying parts of the body increase in size more rapidly than do the shanks, head and other bony parts.

As the process continues, the proportion of fat laid down increases, and finally, at maturity, the properly 'finished' carcase carries the optimum amount of fat and the maximum proportion of flesh to bones, offal and hide. After the zebu reaches maturity, that is, at a weight of about 880 pounds (399 kg.), any increase is due entirely to fat deposition.

The time taken to attain the greatest weight of such a carcase or, in other words, the rate of total body weight gain, has to be considered in the economy of the process. The importance of rapid growth lies in the fact that during an animal's life a certain proportion of the food consumed must be utilized for the mere maintenance of existence before any of it can be converted to flesh.

Thus, the longer the time taken to produce a given weight the more must be the amount of food eaten for maintenance alone, and consequently the greater the expense of obtaining the carcase weight and the less profit from it.

The basic conformation of the animal also helps to determine the proportion of the various components making up the total weight. The cattle of those breeds which have short, fine bones and thick muscles, long, broad backs and square rumps, carry more of the valued flesh and a greater proportion of it in relation to the other constituents of the body than do cattle with long limbs, narrow, flatsided barrels and mean rumps. The deposition of fat in the body follows an ordered course. In the very young animal there is little fat, and what there is of it is very soft, and is concentrated around the kidney and abdominal contents.

As growth and fattening proceeds, a layer of increasing thickness is laid below the skin and over the surface of the muscles, and finally fat is interspersed between the muscle fibres, separating them out and making the meat palatable and tender.

The process is carried through most quickly in the early maturing breeds, and intramuscular fat is deposited at an early age, in contrast to what occurs in the unimproved breeds where the first stage may be prolonged and the third suppressed. A certain amount of fat is desirable but the production of fat merely to increase weight is a gross waste of food.

How rapidly and completely the process takes place depends not only on an abundant, continuous supply of good food but also upon inborn ability. The characters which determine conformation, early maturity and rapidity in gain of body weight are hereditary ones and cannot be brought into effect without a continuous and liberal supply of food of fairly high quality.

The regions in the tropics where the soil and climate allow intensive cultivation of food crops, are populated at a density which these crops just manage to sustain and any enterprise which tends to lessen their direct use as human food would involve a revolutionary change in agricultural practice.

In regions where the soil is less fertile, the num ber of humans relying on it correspondingly decreases but is always heavy enough to forbid the use of crops which can be directly turned to human food being diverted to a process which ultimately results in a loss of calories. Thus it is that beef must for the time being be produced either from by-products of food crops or from pasture and herbage.

Most of the crop waste is needed for work-oxen and milk cows, and so cattle, as beef cattle, must be reared elsewhere: their essential function is the conversion to meat of fodder which can be used in no other way.

In the discussion on pasture in other chapter of this book, the chief differences between that available in the tropics and that produced in other latitudes are described: the restricted periods of growth which nevertheless result in superabundance, followed by the periods of relatively rapid deterioration and long periods of stagnation, are outstanding features in many territories; in others, the arid, infertile nature of the land upon which the pastoralist depends so much is impressive. In either circumstances the period of plenty is too short to allow young stock to complete sufficient growth to carry them to the condition where they might profitably be turned into meat.

Instead they are subjected to periods in which a bare maintenance sufficiency of fodder is available, if all goes well, otherwise they must get along as best they can on a sub-maintenance diet, which results not only in a check in their growth but in an appreciable loss in body weight.

It has been observed that under certain conditions the zebu retains body weight, or loses less than temperate type cattle, but, when adequate feed again permits, the latter type makes the faster gain.

As things are today, meat for the home market must be cheap, and thus the expense of buying or cultivating special feed, where that is possible, or conserving sufficient fodder upon which to fatten, can seldom be incurred.

Similarly, if the meat is to compete in the world market, it must be produced at a price which will absorb the apparently unavoidable loss in weight incurred in marketing the live beast, and the expense of transporting the dressed carcase.

There is a further handicap associated with beef production in the tropics, in that one effect of the high ambient temperature is to decrease appetite and perhaps slow down body metabolism with consequent retardation in liveweight gain, but the cattle of some breeds are less affected in this way than others.

There is, however, some evidence to suggest that the ability to utilise tropical vegetation may be as important as differences in heat tolerance to animals reared in these environments.

The type of animal best suited for these conditions is obviously not the early maturing, short-legged, delicate feeder, but the longer legged, active beast with a lengthy, broad, straight back, the longer well-developed, deep, fleshy hind-quarters, and a barrel associated with a capacious paunch and the ability to deal with large quantities of rather coarse vegetable material; an animal, above all, which is sufficiently unaffected by great heat to allow it to gain weight as quickly as the food supply permits at any season of the year, and one which is of a quiet enough disposition to make handling easy.

Such a beast thus varies materially in essential make-up from that required for stall feeding or for dealing with cultivated crops and lush pasture. In any case it is necessary that the cows are capable of producing and rearing two strong calves in three years.

BEEF CATTLE ON OPEN RANGE

On ranches overhead and handling charges per animal decline as the size of the unit increases, so there is a tendency to maintain the units as large as possible.

The greater the number of cattle composing a unit, the less the attention which can be given to each individual. In breeding operations this means a lower percentage of calving.. A unit of 50 cows to one bull insures higher fertility than one of 100 cows and three bulls at no greater expense.

A unit of 100 cows is as large as is consistent with adequate individual attention: barren cows and shy breeders can be discovered and dealt with, the ageing cows recognized and culled, and the general condition of all easily checked.

Fattening Stock

The essential procedure in the production of good-quality meat is to keep the animal continuously gaining weight. During the process appetite lessens and the animal becomes more and more particular in its choice of food. Thus, when changes in the feed have to be made the quality and palatibility of the new feed should be superior to that of the old.

The tropical grazier can very seldom avoid a seasonal growth check, nor has he a choice of feed to offer, but all grazing ranges vary in quality either naturally or because of introduced improvements, and by taking a little care in the herding it is usually possible to favour the stock at the two periods when it pays most to do so, that is, at weaning and at finishing.

This is possible only when the land is fenced or when the herding is done more conscientiously and more intelligently than usual. Faulkner and Brown have pointed out that one of the most common forms of mismanagement in herding in Africa, where cattle are kraaled at night, is the practice of taking them to graze late in the day and bringing them back too early in the evening.

However, animals eat faster when grazing time is restricted, and if that time is sufficient to allow them to gather the amount of fodder they need, restriction may do no harm.

Fattening or growing cattle should be disturbed as little as possible. Those of a quiet disposition suffer least from any hustling involved in mustering, but others, of which unhandled zebus are notorious, add enormously to the drovers' work and take days to settle down after the rushing and excitement. It is therefore important that such cattle should be given sympathetic handling when very young and become acquainted with the mustering procedure when it can be done quietly, so that, being familiar with it, they will not be unduly upset by the later process, which of necessity may be somewhat rough.

On good feed at the best time of the year, a gain of as much as 2.5 pounds (1.13 kg.) per head may be made each day, but an average daily gain of 0.75 pound (0.34 kg.) over an uninterrupted period of about three-and-a-half years would be a very satisfactory achievement from tropical grazing land.

A more usual rate of gain is that reported from Rhodesia where in an experimental herd it was found that the average gain from birth to the age of 18 months was 0.7 pound (0.32 kg.) for Boran cattle, 0.52 pound (0.24 kg.) for Africander x Borans and 1.0 pound (0.45 kg.) for Hereford x Borans.

In Tangania, lower gains over a similar period are recorded, namely, 0.5 pound (0.23 kg.) per day for pure zebus and as low as 0.39 pound (0.18 kg.) for Hereford x Tangania zebus. In Kenya well-managed Borans made an average daily gain of 0.8 pound (0.36 kg.) till five years old.

What usually happens is that an appreciable gain made after the rains is lost in the dry season and, as a result, slaughter weight is not reached till the stock are four to six years old. In such circumstances not only is the turn-off very low each year (in Queensland, it is nearly 17 per cent and in other territories 10 per cent is considered quite normal, or even good) but the quality of the carcase is low.

Whether it is possible or economically worthwhile to prevent the dry-season loss by fodder conservation or even by crop feeding depends upon local conditions. Assuming that it is possible to make hay or silage, or to grow green fodder or cereal crops, the main considerations are the cost of doing so against the ruling market price of beef, the increment, if any, for quality, and the increased rate of turnover of stock.

In many territories immediate access to a market is not always available and it may therefore be necessary to hold the stock longer than would otherwise be desirable and, in these circumstances to hasten production might be uneconomical. On the other hand, a supply of concentrates or extra fodder over a period of a month or so may allow of the disposal of a beast which would otherwise have to be retained for another year. At the time when a decision has to be made on such a point, the grass and herbage will still be sufficient, if not plentiful, but will almost always, if not invariably, contain a very low proportion of digestible protein.

Although the requirement of that food element for fattening, as opposed to growing, stock is not large, a protein concentrate such as cotton cake or groundnut cake fed at the rate of as little as two pounds (0.91 kg.) per head per day will result in faster gains and better finish.

Low quality roughage such as is had from dry weather grazing and field crop residue can be turned into useful feed by the addition of urea and molassis when molassis is available at an economic price. A mixture of urea, molassis and water in the proportion of 1 : 8 : 8 may be used. Cattle should not be given more than three to four ounces (85-113 g.) daily.

When grazing has been uncontrolled, the best of the feed, growing on low-lying land and around water holes and wells is consumed first and the stock have then to go farther and, farther afield in search of it. An increasing amount of energy is thus expended at the very time when it is most necessary that energy should be conserved.

It is, of course, undesirable that stock should have to walk long distances, and as the water point is the governing factor, the area available for grazing should be considered as that which lies within five miles (8.0 km.) of water.

In estimating the carrying capacity, both water and feed must be taken into account, as the supply of one may put a limit to the usefulness of the other. In the seasons of less extreme heat, cattle will not necessarily water every day if a supply is not at hand; in dry, hot weather they generally drink every day, but some individuals, even then, drink only every second day, but in any circumstances they are reluctant to move from the source farther than is necessary.

The natural tendency of the stock is to feed around the water area until it is cropped bare, that is, within a circumference of about two miles (3.2 km.); beyond that the feed is not evenly utilized unless grazing is controlled or is scarce enough to make everything worth eating; beyond five miles (8.0 km.), grazing is patchy under any circumstances.

Where water may be had at only very long intervals, say at about 20 miles (32.2 km.), on unfenced land under tribal control, such as occurs in many parts of Africa and in Arabia, herds are directed from one water point to another along the lowest-lying land upon which the only herbage grows, until not only are the areas around the points denuded of vegetation but the outlying feed is reduced to the scantiest proportions. Thus serious overstocking, parasitism and soil erosion occur where apparently unlimited space might have been considered an obvious obstacle to such catastrophies.

Presuming that the water supply presents no difficulty, there are so many other varying factors which help to decide the carrying capacity of the land that only a very rough guess can be made until experience teaches what it is.

The annual rainfall gives, perhaps, the most reliable indication, but it is only an indication: the same amount of rain in different places may have very different value according to the nature of the fall, the run off, the holding power of the soil, the prevailing humidity and so on.

There is always a tendency to overstock, which is encouraged by a favourable season or two; a drought follows, or expectation of rain is not fulfilled, and heavy losses occur. In an attempt to recoup, the land is too heavily stocked again as soon as conditions permit and, while the cycle is repeated, the land and the feed decline in quality each year.

The optimum number of cattle is the number which can be carried comfortably through an average year, allowing a margin for adversity and permiting pasture improvement on the lines described in other part of this book, in good years. Where hay, silage or other supplementary foods can be produced, the carrying capacity of the land is increased to an extent corresponding with the size of that reserve.

Breeding Stock

The breeding herd demands somewhat different management from the store and fattening cattle. As long as the grass or herbage is fresh and not wilted, it probably provides complete nourishment, but in the soil of certain localities one or other food salts may be deficient or, more rarely, there may be an undesirable excess of some mineral.

If the nature of the mineral content of the land is not already known, a deficiency of any significance will reveal itself in low fertility of the breeding stock, weakness of the calves and slow growth of the young stock. How such a condition should be treated has been indicated in other chapter of this book.

Even after the feed on the range has passed its prime, it will probably be sufficient for the in-calf cow as long as there is plenty of it and it can be obtained without undue exertion. It is when the grazed fodder is scarce and low in quality that the stress becomes too great, and the growth of the foetus becomes adversely affected in spite of the supplements drained from the dam's body. The condition of the stock should never be allowed to get to such a state, if it can be avoided, for if it does not result in the loss of the greater part of the calf crop, it will at least prevent the realization of its full value, because of the lengthened period necessary for stunted beasts to attain slaughter weight and because the fertility of the cows may also be adversely affected.

The feeding of hay, straw or some such roughage when the grazing is insufficient to maintain condition is most desirable. It is during the last two months of pregnancy that the greatest nutritional need of the foetus occurs.

Some ten pounds (4.5 kg.) per cow each day of any of the common straws, or even coarse grass straw, will be found to be most beneficial: if two pounds (0.91 kg.) of concentrates can be added, all loss of condition will generally be avoided. Urea and molassis fed as described on page 208 may suffice.

These are minimum requirements. When it is possible, as it often is, to grow a sufficient acreage of cereal, the needs of the breeding cows are met from the straw, that of the calves from the grazing afforded at the early stages of the crop, while the offal and the grain itself may be used for finishing when it is profitable to do so.

The feed of a hundred breeding cows could be supplemented at the rate suggested for two months from some ten acres (4.0 hectares) of maize or about 20 acres (8.0 hectares) of sorghum.

The herd of the roving pastoralist obtains supplements from the gleanings of cropped arable land, from the last of the year's growth of herbage in riverine areas, and from browse.

With good feed and active bulls, a calving percentage of about 90 can be attained; should it fall below 70, defective management is indicated, in the absence of disease or abnormal climatic conditions. Such a herd would require at least three bulls for each 100 cows: when feed is sparse, the ground broken and the property unfenced, more bulls are required.

The bulls of some breeds are more virile than others : those of the Mysore type, for instance, may sire a hundred calves or more in a season and do so for some years, while bulls of many African breeds suitable for beef production are usually allowed 40-60 cows, as are those of British breeds.

The indigenous African pastoralist considers a calving percentage of 75 as very satisfactory, but anything from 50 to 80 per cent is considered normal. In the average herd in tropical South America much the same standard is accepted. It is almost impossible to obtain exact figures concerning the rate of herd expansion of itinerant pastoralists. The Nigerian Livestock Mission, after making a survey of the Northern Region, a survey which, at best, could be but superficial, estimated that from every million cattle 52,000 cattle were available for sale each year, or 5.2 per cent.

Their inquiries led them to believe that for each million cattle there were some 300,000 breeding cows and heifers, that the average calving was 40 per cent, that the calf mortality was about 15 per cent, and that of the herd five per cent.

Against these figures they quote others obtained by the administrator of a province as follows: number of breeding cows in each million of cattle, 300,000; average calving percentage, 70; calf mortality, 18 per cent; herd mortality, eight per cent; annual rate of output, 9.2 per cent.

A cow calving in the open generally needs no attention if she is undisturbed. Neither she nor her calf runs the same chance of infection as they would were they in stables, pens or yards.

Difficult and abnormal calving is a rare occurrence in pure-bred zebus but more common in zebu x European cows and in those of some British breeds. All should have regular supervision, but manual interference is not required unless parturition has not been completed in some seven to eight hours.

Even after that length of time if the presentation is seen to be normal, that is, if the fore feet and head of the calf can be seen, it is better not to interfere as long as any progress at all is being made, for the chances of normal parturition is good and the dangers always associated with handling are avoided.

In arid regions breeding should be so arranged that the calves are dropped as soon as the grazing is such that cows can gather sufficient for their needs without covering any great distance. If that is really early in the season, the calf will be grown, weaned and keeping itself on fodder by the time dry-weather conditions are advanced.

If the breeding is uncontrolled, most of the calves will be dropped at that time of the year, but many will not, and serious loss arises from cows dropping weak calves after a period of semi-starvation and when they have to move farther in search of food than the calf can go.

Under good management, in normal years, a free-range herd should have a calf mortality of not more than 5 per cent but, in fact, in many herds where the husbandry is more intensive, it is as high as 10 per cent and often not till it has reached 15 per cent does it attract serious attention. When calves are running with the cows the commonest cause of death is that mentioned above.

When they are separated from their dams during the day, but are not housed, the most frequent causes are, firstly, indigestion resulting from irregular feeding with a superimposed infection of the bowel, secondly, impaction of the stomach with ingested fibrous grass and suchlike material and, thirdly, parasitism.

In regions favoured by a comparatively high rainfall, the bulls should be put to the herd at the time calculated to allow the majority of calvings to take place about a month before the beginning of the rains so that the increased vegetation which follows its onset is available to maintain and perhaps raise the milk yield at a stage when it is still possible to do so and when the calf is old enough to take advantage of it.

The grazing will then probably last long enough to give the calves time to grow and be well weaned before they must meet the full rigours of the next dry season. Only when the bulls are so closely paddocked or confined that they cannot find a maintenance ration for themselves is it necessary to hand feed.

At first roughage will maintain them in condition, but later concentrates should be fed so that when they are required for service they are putting on weight, are well fleshed but not fat, and ready to pay more attention to the cows than to the grazing. Their feeding requires individual attention, each bull getting what his weight and appearance show to be necessary.

On the average, a 1,000-pound (453.6 kg.) indigenous bull will need six to eight pounds (2.7 to 3.6 kg.) of cereal offal in addition to coarse fodder fed to appetite. During the month before service begins up to five pounds (2.3 kg.) of an oil cake and 15 pounds (6.8 kg.) of good-quality hay would be adequate or, if silage is to be had, up to 30 pounds (13.6 kg.) daily may be given with advantage. Such hand-feeding can be justified only when good enough grazing is not available.

Young Stock

Weaned calves are peculiarly liable to infection by the common contagious diseases and parasites, and they should not be exposed to it by confining them in paddocks or yards which have been occupied by the sick or have been fouled by a concentration of mature beasts, as is so commonly done.

It is wise to keep all young stock separate both from the fattening and the breeding herds. Heifers should be drafted into the latter herd when they have grown sufficiently to be bred from; usually that means when it is judged that they have reached a certain weight or have attained a certain age which varies according to the breed and location.

It is customary to castrate all bull calves designed for meat although it has been unquestionably demonstrated that uncastrated males grow faster and require less feed per unit of gain. The reason why castration is favoured where quality meat is in demand is because much of the weight of the uncastrated animal is put on where it is least wanted, that is, on the neck and breast, which, along with excessive darkness in colour and the somewhat strong flavour associated with bull flesh, lowers the carcase quality.

The carcases of young bulls do not have these defects to any great degree, and as there is a smaller proportion of fat and a higher proportion of edible meat on the carcases of bulls than on those of bullocks, it would appear that no advantage is to be gained by castrating bulls required for local tropical meat markets. Castration, however, has other uses: it prevents uncontrolled breeding, and the disturbance caused in the herd by fractious young bulls.

The longer castration is delayed, the more likely are these undesirable features to appear, but delayed castration, say to six months, does give a better opportunity to select breeding bulls for rate of growth and general performance. When it is work and not beef that is required castration may be delayed till the animal is 2-21 years old. The younger the age at which the operation is performed, the easier it is to do and the less shock it is to the animal.

Heifers which are grown for meat are castrated or 'spayed' at an older age: it is said to hasten their fattening and improve the quality of their carcase, but at least one controlled experiment has shown that no advantage results. It does, of course, prevent them breeding. The operation is more difficult to do than castration of the male, but is a simple one to the expert.

Where the danger of predators is at all serious, dehorning is not practised and the stock are left with their powers of defence unimpaired. Where they are to be mustered and handled often or held in confined conditions for any length of time, dehorning prevents injury to the hides and promotes quietness in the herd. Dehorned, yard-fed cattle, all other things being equal, gain weight quicker and more economically than horned ones. Dehorning methods are discussed in the previous section.

It is not practicable to dehorn each calf at an early age in the freerange herd, but the operation should, if possible, be done when the calf is between three and five months old, by which time the horn will have grown to a handy size but is still small enough to be dealt with quickly and easily.

Dehorning can be carried out at any age, but as the animal grows, the operation becomes longer, bloodier and more painful. It consists of removal of the horn and not less than a quarter of an inch (0.63 cm.) of the surrounding skin with special cutters or a sharp fine saw. It should be done at a period of the year when flies are least troublesome.

When the calves are weaned they must be completely removed from both sight and sound of their dams, otherwise neither settle down quickly. If the separation is complete, a day or two suffices to accustom both to the change.

This must not be made, however, until the calves are well able to fend for themselves and, if possible, when feed is abundant or, at least, when it is easily come by. Their digestive powers are not at first fully developed and must not be overtaxed, so they should have the best grazing procurable, not only for that reason but because no better use can be made of good food and because it will build them up for their first dry-weather season.

When hay or silage is available to supplement the grazing and browsing during the period of scarcity, the young stock should have priority in its use: so much benefit do they derive from such timely feed that it is worth considerable expense and trouble to arrange it. In any circumstances a supplement of common salt and any other mineral known to be necessary should be at their disposal.

In hot weather shade should be provided: it has been shown to result in an appreciable increased rate of growth. Cattle of indigenous tropical breeds use shelter in the heat of the day very much less than do those of exotic breeds or their crosses. Calves shelter more than do older animals, and young-stock more than adults.

Probably the most effective form of shade is that afforded by a well leafed tree and it is worth trouble and expense to arrange for such shade whenever possible.

If artificial shade is to be provided, the best overhead cover appears to be the traditional thatching of straw or reeds, but, because of the need of annual repair or renewal it is, in the end, rather expensive. Corrugated iron, which is so often used, has been shown by Kelly, Bond and Ittner to be particularly unsuitable unless it is overlaid with such material as straw or grass.

They recommend that whatever material is used, the structure should be oriented east-west and should be ten feet (3.0 metres) high.

ENCLOSURES

Without careful day and night herding, such as only the most conscientious attention can ensure and which it is not practicable to give on a ranch, a large number of cattle cannot be. managed in detail unless the land is enclosed. Fencing has been the forerunner of both land and stock improvement everywhere.

A settled cattle population, divided into its component parts in convenient-sized units not only allows of a variation in management to suit each unit but permits the best use of the land and water resources.

In the absence of fences, water points tend to be crowded, the best feeding areas are overgrazed, young calves are walked off their feet, the in-calf cow starved, and the barren cow preserved. On unfenced, private land, neighbours' stock get mixed and compete for the best feed.

On public land where the control of grazing is not vigorously exercised and especially where there are communal rights of grazing, competition for the largest share is intense and continuous and, almost invariably, laws designed to preserve its usefulness are ignored, circumvented or found to be unenforceable, with the result that the land is abused and its fertility destroyed.

It would be quite impracticable to fence much of the grazing land in many parts of the tropics, in fact today practically the only fenced land in Asia and Africa is that owned or leased by Europeans or by governments. That does not mean that it is not desirable to enclose grazing land, but only that for the time being it cannot be done.

This can be said in support of the complete free range-it saves heavy capital outlay, less labour is required, and uncontrolled stock will always find the earliest growth and the best feed.

It is generally desirable to have stock divided into the following categories: weaned calves; young stock, comprising both bullock and heifers; breeding cows; fattening stock at different stages of growth. How many subdivisions are necessary and how many units of each it is advisable to have, depends, of course, upon the total number of stock and what area per head the quality of the grazing allows.

The paddocks for the weaned calves should be situated on the best land to which there is easy access : that will almost invariably be in the vicinity of the steading or headquarters. The breeding herd, requiring the supervision it does, should also be situated more or less at hand, and paddocks for the other stock may be located beyond. One or two small enclosures at home are required for the sick and for bulls, and a larger one for mustered fat-stock.

The water supply may dictate where and how the paddocks will be situated, but fenced land when judiciously located should permit the installation of a central water system to supply all main paddocks and yards. A paddock without a water supply is of very limited value.

Immense areas of potential grazing are devoid of use because, as they are at present, drinking water is not obtainable: much of it is good enough to justify the expense of developing a water supply, as has been demonstrated in Uganda in a small way, and in Australia in a more impressive manner, but there still remains a vast acreage upon which the feed is too poor to pay for even a moderate outlay, or the water is so inaccessible that the cost of development is not justified.

As a supplement to springs and semi-permanent stretches of surface water the construction of wells is usually demanded, or tubes may be sunk to a natural underground reservoir, or a supply may be piped from a river or dam.

Wells are expensive to build and their output is often too small but they permit of water being drawn by hand, the only method practicable when the well is outlying and used by unsophisticated people. No supervision of economy is required but stock may go short of their need. Under other conditions both well and tube supply is raised by mechanical power.

Windmills have proved economically satisfactory where air currents are dependable in duration and strength. In Asia, the lift by bullock traction through the Persian wheel or the water-bag has stood the test of centuries and is likely to continue there, as it does where it has been introduced in Africa. In all circumstances a storage tank and drinking trough are desirable if not essential.

A stout circular, concrete tank some four feet (1.22 m.) high with an adjustable spill-over to a surrounding trough about one foot (30 cm.) high and just broad enough to allow a full-sized ox to drink from has been found satisfactory.

Dams generally offer the best means of water conservation. Those of any pretension should be designed if not constructed by a specialist, but useful-sized dams can be made without his help. The size should be related to the requirements of the stock rather than to the total amount of water that a catchment area will provide.

The most common mistake made in the siting is to locate it in a valley where the deep sides funnel the water down under enormous pressure, which a dam of only exceptional strength will withstand. It is better to have it located in a shallow valley, not necessarily where it will catch all the run, but where just sufficient water will be trapped and the rest easily and surely diverted.

If a low barrage is built at a suitable distance above it, it will help to prevent rapid silting. An excavated basin makes sure that the impounded water will be held, and supplies the material of which the actual dam is constructed.

Both the dam and its surrounds, unless they are exceptionally well grassed, should be fenced off, and the water for the stock piped to a trough at a lower level, the pipe being laid under the dam before its construction.

Perennial springs and streams can be likewise developed so that they feed a dam, or they may be excavated, or pooled, to expand their usefulness.

At all sources of water it is most important that access for stock is made as easy as possible and that there is accommodation for a sufficient number of cattle to drink in an orderly fashion at the same time. A crowded mob of struggling cattle results in injury and sometimes death of stock, deterioration of the approaches and of the surrounding land, and contamination if not destruction of the water supply itself.

When practicable, the water should be piped into troughs, but that usually necessitates a pump of one kind or another. When troughs cannot be used, a substantial approach for stock should be laid up to, and into the water if it is shallow, a sufficient frontage being allowed. Other approaches should be fenced off. Where the banks are deep, the approach should be cut through them and the greatest care taken to see that the frontage provided is adequate.

The amount of water required should be calculated at a minimum of ten gallons (45.5 litres) per head each day, plus that likely to be lost through evaporation and seepage.

Fencing

Fences may be 'live', that is formed by growing trees or bushes, or they may be of wire, timber or stone.

Live fencing requires the least capital outlay, but, of course, it can be made only when conditions permit the growth of a suitable plant. It takes time to establish—often a matter of years—and it may never be quite reliable because of breaks in continuity which, even in the best fences, occur from time to time and are not always easy to detect: moreover, many of the tropical shrubs trained to make satisfactory fences in their early years of growth, in time become weak at the base, overtopped and irregular.

Conditions concerning the construction and upkeep of live fences are so variable that no general guidance can be offered. The government forestry or agricultural department should be consulted when help is required.

Wire fencing permits of a considerable choice in cost, both wire and supporting uprights being obtainable in various grades of quality, and the amount to be used being to some extent a matter of discretion.

The expense of the labour in erecting inferior material is the same as that incurred when the best is used: it is considerable under any circumstances, and as the shorter life of the cheaper material necessitates relaying more frequently, it proves in the end to be the more expensive.

A four-strand fence of long-set 14 gauge barbed wire or two strands of number 8 gauge plain wire and two strands of barbed wire suffices for cattle of European breeds and their crosses and, in fact, it can be relied upon to hold cattle of most breeds, if the strands are properly spaced.

The cattle of some indigenous breeds which have been accustomed to free range and are unused to handling cannot always be held by it and an extra strand may be required. Satisfactory fences may be erected of three strands of wire where quiet cattle only are to be contained. Barbed wire undeniably is a menace to hides and often the cause of serious injury to stock when not properly laid but it undoubtedly has the power of holding a press of cattle.

Its ill effects may be mitigated if it is not used for the upper strands of the fence and if it is entirely replaced by smooth wire where ganging and milling are likely to occur as in the vicinity of crushes and in marshalling yards. A beast which is attempting to leap a fence is not dissuaded by barbed wire but it may be badly damaged by it: the beast which tries to force its way through a fence nearly always does so by getting down to it and is discouraged by the barbs.

For ordinary fencing the most useful height is 56 inches (142.2 cm.) with the lowest strand 18 inches (45.7 cm.) from the ground, the next 12 inches (30.5 cm.) from it, the third 12 inches (30.5 cm.) above the second and the top strand 14 inches (35.6 cm.) above the third. If three strands are used, the spacing from the ground upwards should be, 20 (50.8), 16 (40.6) and 18 inches (45.7 cm.), the fence being two inches (5.1 cm.) lower than a four-strand one.

For smaller, quieter stock, five strands eight inches (20.3 cm.)

apart are generally used. Variations to suit local ground peculiarities are, of course, necessary.

The staunchness of the fence, and, in fact, its whole life depends upon the reliability of the supporting uprights or straining posts. These must be spaced and set in the ground in such a manner that they will not only carry the weight of the wire throughout its life but will also sustain at least the normal stress exerted by a weight of cattle.

Posts are normally required from 15 to 20 feet (4.6 to 6.1 metres) apart, or at about 45 feet (13.7 m.) with four droppers between. Their length should be at least six-and-a-half feet (2.0 metres) so that they may be let one foot ten inches (56 cm.) into light soil and somewhat less in ground offering a more secure base.

The corner posts, taking extra weight, must be particularly staunch and should be let at least three feet (0.92 metres) into the ground: they should be securely braced. If there is a long stretch between corners, well-braced supporting posts should be erected at intervals to reinforce the ordinary uprights.

A further useful precaution is to brace two or three of the posts on either side of the corner ones. Special non-vulnerable timber must be used where ordinary wood is liable to attack by termites. Stout metal or reinforced concrete uprights are the most satisfactory but expense usually precludes their use.

On small enclosures, stone walling is a possible alternative to wire fencing, where the material is at hand and labour is not too expensive. A breadth of 24 inches (0.61 metres) is required for a sound base, while the top should be 12 inches (0.31 metres) broad to be effective. The height may be as low as four feet (1.2 metres).

It has been found that 33 yards (30.2 metres) of such a wall should be built by ten African labourers in about six hours and that the same number of men are required to prepare the stone. Such walls must not be allowed to get into a stage of disrepair: stones get displaced, subsidence causes collapse, flaws in building appear; if corrected immediately, the work involved is usually negligible, but if reconstruction is unduly delayed, one gap promotes another and damage gets out of hand.

Where regular, periodic supervision can be given, electric fencing is satisfactory and comparatively inexpensive.

Conventional hung gates are not often satisfactory; primitive people are unused to them and abuse them, more sophisticated people neglect them and in both cases the hinged gate is seldom either properly open or securely closed. The fault usually lies in the way the gate has been hung and in the nature of the lock. Both timber and metal gates should be very strongly constructed, the gate posts should be immovably fixed in the ground, the gates should be hung so that they tend to swing closed rather than open and they should be fitted with a simple, strong, automatic locking device, of which there are many, which cannot be opened by stock.

The breadth of the gateway should be at least 12 feet (3.7 metres), but if large mobs of cattle are to be handled, a space at least twice that breadth should be provided. Such wide gateways can be closed only by some such device as a wire gate which is made by three or four strands of wire strung between battens with two 'droppers' placed equidistant from the extremities to keep the wire properly spaced. The battens are secured to gate-posts so that they can be easily removed or can be made taut by a lever action.

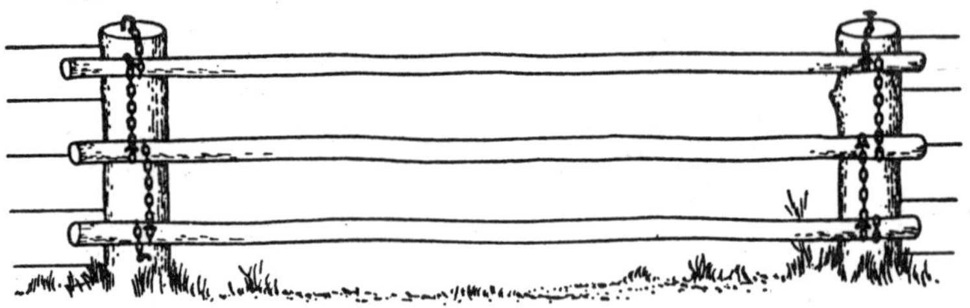

Figure 7.6: Pole and chain gate.

For a 12-foot (3.7 metre) span Robertson describes an inexpensive gate made with poles connected by a chain passing round each pole and fixed to it by a staple. The 'hinged' end is attached to the top and bottom of the gate-post by a stout staple.

The other end, when in the closed position, is hung from a large nail or hook inserted into the top of the gate-post from which the gate at this end hangs free or is fastened over a nail at the base of the upright. The poles must be long enough to overhang the gate-posts as shown in Figure elsewhere in this chapter.

Handling Yards

Such yards are necessary when an appreciable number of cattle have to be dealt with. Their number and size will depend upon the head of stock to be handled but they usually include assembly, holding and drafting yards, as well as a work yard where a crush and other facilities are situated.

Whatever the design, there are certain considerations applicable to all. The most important of these are that the yards are located on ground which can best withstand constant wear and which under all circumstances is well drained; that the containing fences are undoubtedly strong enough to withstand the shock of bunched cattle attempting to break through or press along them, and high enough to prevent a beast from jumping over-for large, active zebus, that is, five feet six inches (1.68 metres) in the main yards and six feet (1.83 metres) at the approaches and around the crush; and that the fences must be kept free from projections which might cause damage to the stock or workers.

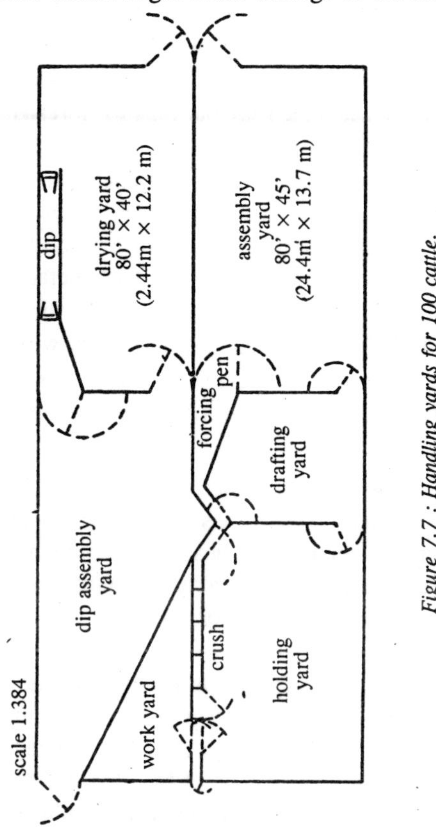

Figure 7.7 : Handling yards for 100 cattle.

Wire fencing is not suitable for such yards; something which is strong and is seen to be strong is needed. Substantial timber such as eight-inch by three-inch (20.3 × 7.6 cm.) split rails or six-inch by two-inch (15.2 × 5.1 cm.) sawn timber is satisfactory in most circumstances.

Tubular metal is the ideal material, if cost is of secondary consideration. Stone walls or other such solid structures are objectionable not only because they make the yards hot but because they hinder the worker's movement and obstruct his sight. When six-inch (15.2 cm.) timber is used, four bars spaced 10, 9, 9, and 14 inches (25.4, 22.9, 22.9 and 35.6 cm.) from the ground upwards is satisfactory.

They should be supported by uprights not more than seven feet (2.13 metres) apart, and set at least two-anda-half feet (0.76 metres) in the ground.

The yards should be located where the natural features of the land promote the movement of the cattle in the direction required, and they should, in any case, be constructed where water is at hand.

Very often the location of the yards will determine their design, but wherever they are situated their lay-out calls for foresight and ingenuity to facilitate work and protect the stock.

The design illustrated in figure elsewhere in this chapter is a simple one which has been found satisfactory when numbers of cattle have to be passed through quickly, for instance for inoculation or branding, and for the more intimate handling needed in such operations as dehorning.

The advantage of having the dip at hand, and the facilities offered for drafting and cutting out groups from the herd will be appreciated.

Several useful plans of handling yards designed by experienced men in Australia are described and illustrated in detail by Beattie.

Cattle Crushes

The construction of such a useful tool as a cattle crush deserves more consideration than it often gets. The essentials are that it should be strong enough to contain any beast put in it, that it should be so constructed that damage to the animal is prevented, and so designed that it offers the necessary facilities for handling with the minimum of labour.

A common fault in design is the provision of too much room for the individual animal: this encourages struggling, attempts to jump retáining barriers and to crowd into already occupied standings. The most common fault is the provision of too great width at ground level in the crush. Figure elsewhere in this chapter illustrates dimensions which have been found suitable for large zebus.

Tubular metal-old boiler tubing is a cheap and satisfactory form - and concrete are the best materials for construction, but old railway lines and sleepers adequately take the place of tubular metal and, in the absence of both, tough timber, either sawn or in the form of rough-dressed poles, is satisfactory.

The work to be done in crushes may take but a few seconds for each animal or it may occupy some considerable time. In the former circumstances a holding arrangement is needed which can be quickly brought into operation or as quickly removed so that the flow of cattle is not delayed: it must therefore be both effective and simple.

As it is often desirable that several beasts be held simultaneouslyfive is a convenient number-the race or approach to the crush should be of sufficient length to accommodate them. The control may be a drop gate which, when down, separates one animal from another at each pair of supporting uprights, or the animal may be confined by a pair of poles of about four-inch (10.2 cm.) diameter, or even by a single pole, slipped through generous-sized brackets on opposite uprights.

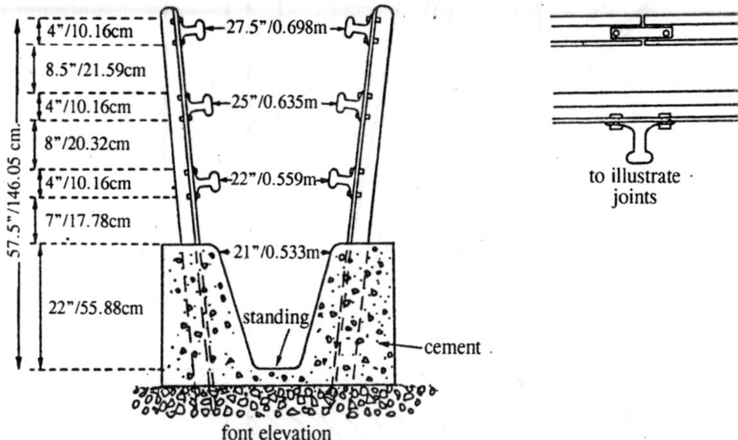

Figure 7.8: Cattle crush: front elevation.

This arrangement effectively restrains animals for the short time required to receive an inoculation, for instance. When necessary, stricter control can be effected by securing the head to the upper bar of the side of the crush.

Work which entails more handling should be done in the crush proper, that is the foremost section, which it should be noted has a length of no more than six feet three inches (1.9 metres). The main feature of the crush is a front retaining gate in the shape of a yoke with which to secure the animal's head.

At least one side of the crush takes the form of a hinged gate to allow easy access to any part of the animal, and through which the animal is released. The side gate is swung from the back upright : it can be made so that the upper and lower half work independently or as one. It is generally agreed that the former arrangement is the more satisfactory. The retaining yoke is constructed of two pieces of two by four inch (5.1 × 10.2 cm.) tough timber, one of which is fixed one-third of the way across the front gate while the other, moving on a bolt at its lower end, is levered into action by a similar piece of timber let through the upright as shown in Figure elsewhere in this chapter.

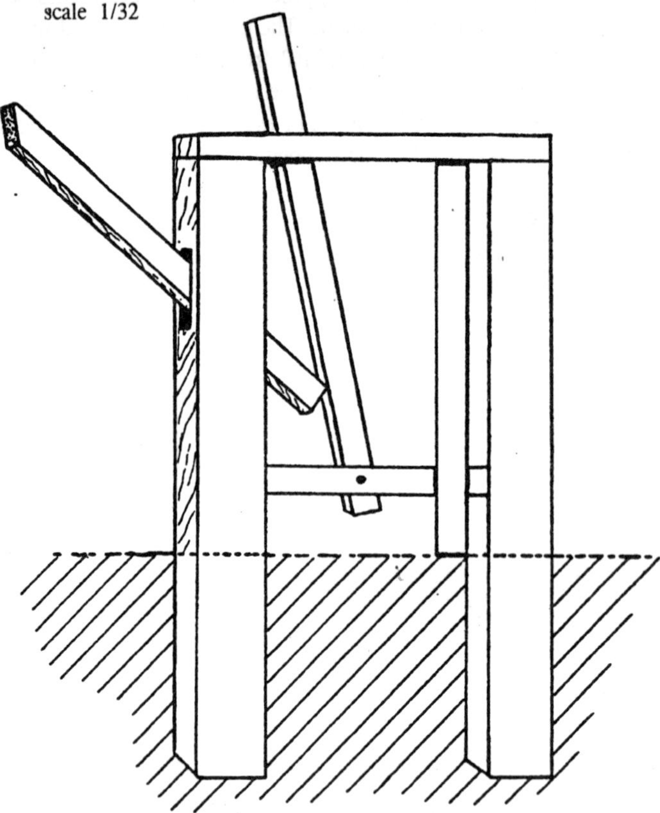

Figure 7.9: Cattle crush: retaining yoke.

When the yoke is closed, the lever is held horizontally by a metal tongue which swings into position from a bolt over the mortising. Where large numbers of cattle of different sizes have to be dealt with, both uprights are movable around bolts at ground level and are brought into play by overhead, jointed, metal arms, acting scissor-wise, controlled by a metal shaft convenient to the operator standing at the side.

Beattie thus describes a very serviceable yoke used at the Veterinary Research Station at Glenfield, New South Wales.

'The yoke, an inverted hollow" Y" shape, should be of tube at least one-and-one-half inch (3.8 cm.) in diameter, spread at the ends to guide the neck of the animal into the narrow section, and further extended to allow the ends to run in the hardwood channels on either side of the bail.

It is raised and lowered by a cord taken over a pulley attached above the bail to either the roof or, in the case of an outside construction, to an extra board built specially to take it. The yoke cannot pull out as the spread ends of the piping are held when they meet the cross pieces at the tops of the side channels.

It can be dropped quickly and is quite firm without the aid of levers, ropes or bolts. A suggestion is that yokes of two or three widths be kept so that large beasts can be accommodated without inconvenience through having a small yoke jammed down on them.

In the cattle crush at Glenfield the inverted "Y" yoke is a permanent fixture, the cattle being released through the side gates, but in certain circumstances it would probably be an advantage to fit the bail to a gate, through which the animals would be released.'

The whole length of the crush should be floored with concrete and, when practicable, the concrete should be built to a height of two feet (0.61 metres) on either side of the race to form a supporting base for the uprights, a protecting wall against injury to the feet and legs of the stock, and a platform for the workers-no other material will so well withstand the continuous heavy, damaging wear to which this part of the crush is subjected.

TRANSPORT OF TRADE STOCK

There are few beef-raising areas in the tropics which are served with adequate transport facilities and seldom has the stock-owner a choice of how his cattle should be conveyed to market; they might, however, go by one of four ways, namely, on the hoof, by road, by rail or by air transport.

Movement of Foot

Cattle which are almost continually on the move in search of food should not be much inconvenienced by moving in one direction for the same distance as they do on their feeding range each day.

Thus it might be thought that droving would in all circumstances be the most economical way of travelling. For short distances, it undoubtedly is, but over long distances such a categorical statement about it cannot be made because many considerations are involved. Droving may well be the least expensive method when an adequate supply of food and water can be arranged at suitable points along the route or when a herd can be watered night and morning and where during the interval it can move at leisure from one post to another while grazing on the way.

These happy conditions are seldom found, at least after the herd has joined the main route. On these tracks feed soon becomes scarce, and as one drove follows another, it well-nigh disappears. For a time it can be obtained by skirting the tracks but, on land where there is no objection to roving, it eventually lies too far out to permit of its use if the cattle are to make a reasonable rate of progress to their destination.

More often, the owners of the land in the vicinity of the track will not permit grazing. As well as these two sound reasons why trade cattle must be kept to a well-defined track, there is one other, that is, the danger that the wandering cattle may spread contagious disease in the country through which they pass. As long as they are kept to defined routes, it is possible for the authorities to check their health from time to time, to control disease if it should appear and to give protection to those in danger.

As a controlled, danger-free, unimpeded passage of trade cattle through a country may confer as much benefit to the general community as it does to the owners and because, at any rate, they are often a necessity, trade routes are usually the concern of governments.

Ordinarily it is the location of water which decides how the route will run and it decides, too, the rate at which the stock will move along it. Where watering points are plentiful, movement is at the discretion of the drover but, where they are located at long intervals, he has no choice but to cover the distance before extreme thirst disseminates his stock.

To travel under such stress, results in serious loss of body weight which, on the African West Coast, for instance, has at times been estimated to be as much as 50 per cent. There, as in other parts of Africa, cattle which have successfully traversed country where scarcity of water is the hazard, have to face another danger in the form of a 'tsetse fly belt' through which, if they are not protected, they must pass quickly.

Even under more favourable conditions in other parts of the world, shrinkage of up to 20 per cent of body weight is common, and 10 to 15 per cent is accepted as normal. Only mature cattle are fit to undergo long journeys on foot under ordinary commercial conditions. Indigenous cattle are well served when water points are located 12 to 15 miles (19.3 to 24.1 km.) apart but in many territories that is not possible and intervals of over 20 miles (32.2 km.) have to be accepted; in some localities a two-day journey may separate them.

If fat cattle are not to suffer serious shrinkage, water should be available at intervals of five miles (8 km.) and they should move by road only where there are facilities for grazing; this applies particularly to cattle of temperate breeds and crossbreds.

The quantity of drinking water required by stock is discussed in other chapter of this book. The supply of food offers almost as great a problem as that of water in certain seasons of the year.

In localities where a satisfactory fodder reserve can be accumulated at each post for the supply of cattle passing through, there is usually enough along the track to meet essential requirements and there is no urgent necessity for it at the post; on the other hand, it is difficult to obtain and has generally to be transported at considerable cost to halting posts located in areas where a supply is essential.

Under such conditions, 10 pounds (4.5 kg.) of rough forage per head allows the trade route to be kept open but does nothing to prevent heavy shrinkage.

The loss from death which may be expected on a given route is unpredictable but depends chiefly on the prevailing cattle diseases and the weather, and upon what facilities are afforded to withstand them. Where beef is acceptable in any condition, much of the loss to the owner is avoided by emergency slaughter.

Transport on hoof entails loss from shrinkage, loss from death as well as drovers' pay, taxes, tolls and extortions.

Rail Transport

Over long distances rail transport effects saving on maintenance and handling charges, tolls and other incidental expenses as well as from shrinkage and deaths. Loss from shrinkage under the worst conditions may be as high as 10 per cent but should normally range between three to five per cent on a journey of up to 72 hours.

That does not, of course, take into account losses arising before entrainment. Loss from bruising may be considerable. McManus observed that in Kenya it resulted in down-grading of some seven to eight per cent of carcases and that it could be greatly reduced as the following data indicate:

	Bruising observed	*Not observed*	*Total*	*Percentage with Br.*
Homed cattle	271	378	788	41.8
Polled cattle	28	111		20.1
Exotic Cattle	176	118	685	59.9
Indigenous cattle	152	239		38.9
Bedding used	82	156	915	34.5
Bedding not used	454	223		67.0
Sexes mixed	78	206	915	27.5
Sexes not mixed	83	548		13.2

Railway trucks which are to carry livestock should invariably be roofed, provide for free passage of air and have battened or nonslip floors. If they are not designed for end-to-end loading on the United States pattern they should have movable partitions to facilitate side loading and to help cattle to keep on their feet while the train is moving.

Cattle should be packed in a truck close enough to prevent them being thrown about during transit. They should be grouped so that those of approximately the same size and condition are trucked together. Horned cattle should be secured by the horns to the truck side; they should stand head to tail so that alternate animals face in the opposite direction.

The stock should be watered every 24-27 hours, and, when practicable, should also be fed at these intervals. On journeys of more than three days' duration, off loading for feed and rest should be compulsory, not only for humane reasons but to prevent undue loss.

At the railhead, arrangements must be made for watering, feeding and resting the stock before subjection to the rigours of a train journey under tropical climatic conditions. If grazing is not adequate and fairly easily accessible, a supply of fodder must be available for hand feeding. A minimum of 20 pounds (9.1 kg.) per head per day is desirable.

The water supply should provide at least five gallons (22.7 litres) per head each day and should be delivered from troughs or other receptacles which will withstand repeated stress and wear on their approaches and surroundings.

Where a large number of cattle are to be handled, a reception yard with a crush and sorting facilities should be constructed of a size suitable for the number of stock expected. The sorting pen should lead to small pens just big enough to hold a truckful of cattle, and those to loading pens which in turn should lead by a ramp, with a gradient of not more than 1 : 2 to a loading platform on a level with the floor of the truck. The sides of the holding and loading pens, of the ramp and of the platform should be very substantially built.

Discarded rails or sleepers make very suitable material, but split timber or two by six-inch (5.1 x 15.2 cm.) boarding is very satisfactory. If they are constructed with spacing not more than five inches (12.7 cm.) between the lowest rails, they will contain small stock as well as cattle. The height should be 54 inches (1.37 metres). Animals should not be loaded more than one hour before the journey commences.

Road Transport

Transport by road in most tropical countries is as yet of very restricted use because of the lack of permanent, all-weather roads. Tracks and unsurfaced roads are not only liable to be blocked but occasion such wear on the vehicles that the cost of running them for livestock transport becomes prohibitive; moreover, the cattle are liable to bruising or more serious damage during transit.

None of these disadvantages holds when an all-weather road system is available. The convenience of loading at the farm and direct transit to the destination, the absence of repeated handling and the disturbance associated with it, and the consequent avoidance of serious shrinkage, outweigh the disadvantages, namely, that the journey takes longer and that the cost per mile is somewhat greater than it is by rail.

Air Transport

This method of carrying live beef cattle is too expensive to allow of its use anywhere to any great extent at present. The value of the quick transit it affords is greatly offset by the journey it is usually necessary to make on foot before emplaning and often the one of considerable length after deplaning.

It is, however, under certain conditions, an economic proposition as a means of conveying beef carcases, even though it entails the erection of an abattoir and refrigerator on the airstrip, with all their accessories, and the supply of specialized labour.

OTHER METHODS OF BEEF CATTLE HUSBANDRY

The salient features of an establishment devoted to beef production outlined above refer to settlements where intensive husbandry finds no place. It is, for all practical purposes, the only organized form of husbandry followed in the tropics with beef production as its end. But, as we have noted, the beef supply, for the majority of tropical peoples, comes from the stock of the roving pastoralists and nomads to whom it is an inevitable and, to them, in some ways a regrettable by-product of their other activities.

Their main concern is to multiply the number of their cattle, to prolong the life of each beast and to keep them all in good condition, not for the purpose of sale, but with the object of increasing their chance of survival.

Under present conditions losses are no longer sustained from disease and warlike activities on a scale in any way comparable to what it used to be, and as a result, these cattle-owners to whom the effects of overstocking are becoming glaringly apparent, are now prepared to off load when the facilities for doing so are at hand.

The erection in stock-rearing districts of abattoirs which are prepared to take the stock as it is offered, and the location of near-by markets of suitable consumer goods as an inducement to acquire cash, are promoting a change of outlook which should be beneficial to all concerned.

As yet, however, beef production in these countries is no more a planned industry than it used to be, but the first stages of development are appearing in the form of land enclosure where some of the cattle to be offered for sale to the abattoirs are held for a period to gain weight, if not to fatten.

The practice, so common in the temperate zones, of rearing cattle on poor land to be fattened elsewhere on cultivated crops, is carried out in a relatively small way in Southern Rhodesia, in the Sudan, in Australia and the Americas, to cater for a market which can pay the cost; but, for the reasons already given, it is a system which is unlikely to come into extensive use.

Stall Feeding

Stall feeding of fattening cattle demands settled conditions of husbandry, a plentiful supply of suitable food and an ambient temperature which will not so reduce the animals' appetite as to make the intake of food inadequate. There would appear to be no place for the intensive rearing of young stock as is carried out in the temperate zones.

There, by feeding concentrates alone, or with a minimum of roughage, a body weight of over 1000 lb. (453.6 kg.) is attained at just under one year old. The consumption of concentrates produces less redundant body heat than does that of roughages but the practice is usually uneconomic in the tropics where there is no surplus grain for cattle feed. In many tropical countries stall feeding is never attempted during the hot season but is successfully followed at other times of the year.

The consumption of large quantities of food above the requirements of maintenance and normal activities, which the fattening process entails, generates an embarrassing excess of body heat. The first requirement of stall-fed animals, therefore, is that they should be protected against climatic heat as far as possible and afforded every facility to lose body heat; that means, in practice, that they must be supplied with shelter designed to protect them against the direct rays of the sun and its reflected heat, and to promote the free circulation of air around them.

The indigenous stock-owner, fattening one or two beasts, houses them in a building very much the same as the one he himself lives inthe circular walls of mud, the roof of thatch, some small space at the eaves for ventilation, no windows, but a permanently open doorway.

There is something to be said in favour of this primitive structure; it discourages flies, it effectively excludes direct exposure to the sun and reduces the inflow of hot air, it is conducive to rest, and while it restrains the animal, it allows movement sufficient for comfort.

Its chief defect is that by stopping practically all air currents, it does not permit advantage to be taken of the very considerable relief they afford: loss of heat by convection is particularly important to the fattening animal.

The standing is insanitary but accumulation of manure is an important sideline of the business and probably no other form of stable can so effectively promote the collection and decomposition of excrement and litter. A fattening ox produces about four tons (4104 kg.) of solid manure and about one and a half (1539 kg.) of liquid manure per annum.

The associated process of fermentation does nothing to lessen the temperature within the building and the evaporation of urine increases the humidity of the atmosphere making it unsuitable for the fat beast or for the growing animal.

The defects can be corrected by redesigning the wall so as to permit the free passage of air at all levels. Ventilating spaces can be opened at appropriate heights or, if the wall is made of brick, by leaving sufficient spaces it can be constructed in the form of a screen rather than as a solid wall. When timber is available, rough uprights spaced a few inches from each other can be substituted for the mud wall.

Such walls are satisfactory in a humid climate where there is no great difference in the temperature throughout the 24 hours; where, however, there is a great variation between the maximum and minimum daily temperatures, the solid walls are less objectionable than they might at first appear to be and are often preferable to the semiopen ones. In any case, the roof, preferably a thatched one, should amply overlay the walls, that is, to a distance of at least two feet (0.61 metres).

The most economical size of such a building is one with a diameter of 12 feet (3.7 metres) which will comfortably hold two mature bullocks of the larger breeds. If built of mud or sun-dried bricks, the wall should be eight feet (2.4 metres) high and at least nine inches (22.9 cm.) thick.

Where feeding is to be undertaken in a bigger way, the individual housing described above is not suitable as it involves too much labour and too much housing construction. The best accommodation is afforded by a yard and shelter.

The yard should be fenced in the same way as the one recommended for handling enclosures on page 220, and the roofed shelter, oriented east and west, should be situated in the centre of the yard so that it offers shade at all times of the day. Water and feeding troughs should be placed below the shade and not around the yard.

An area of 80 square feet (7.5 square metres) per beast is generally allowed and, in the tropics, 40 square feet (3.7 square metres) of shade is desirable.

SELECTION OF CATTLE

The cattle of unimproved breeds do not respond to feeding, no matter how good it may be, in the same way as do those of the specialized beef breeds, and a comparable turnover is not to be expected from them; but, on the other hand, cattle of the improved-breeds may not thrive under tropical conditions and their potential productivity may be of no account.

When selecting cattle for stall feeding, therefore, the prevailing climatic conditions must first be considered from the various aspects discussed in other chapter of this book; it will then be decided if the breed to be used must be an indigenous one or if the conditions are temperate enough to permit the choice of the exotic beef breeds or their crosses.

That having been decided, a choice has still to be made between the breeds of either category. In some instances that would present no difficulty: for instance, in Kenya one would not hesitate to use Boran rather than Nandi or, if one had to make a choice of Indian breeds one would not select the Red Sindhi or Sahiwal for fattening. Often the choice is not as easily made because the difference between the breeds is not so clear cut: the difference may be there, a difference in growth rate and temperament which make some cattle distinctly more suitable for confinement and fattening than others, but it has perhaps been masked by the conditions to which the cattle have been subjected.

Many native breeds have unsuspected inherent possibilities which are discovered only when they are raised in favourable conditions. We have seen White Fulani cattle, when well fed, reach slaughter weight before they are three years old, that is half the time taken by cattle of the same breed kept under standard country conditions.

Vorster records how indigenous cattle which, under native management were nine to ten years old before they reached slaughter weight, when fed a fattening ration from weaning reached 1,000 pounds (453.6 kg.) live-weight at two-and-a-half years of age. Nevertheless, in normal circumstances, the local cattle of the breed with the best reputation for fattening, and those of them bearing the characters described on page 204, should be chosen.

It is, in any case, inadvisable to attempt to fatten flat-sided, narrowchested, long-legged beasts in poor condition.

In the general course of events, the fattening period must be short, that is, no more than a few months, and so animals which will give the biggest return within that time must be chosen: three-quartergrown beasts, in fair but thriving condition which have been on no better keep than the average coarse grazing, best answer that requirement.

When a longer period of feeding can be allowed and market conditions are favourable, yearling bullocks may be brought on quickly enough to give a worthwhile return.

If cattle which are strangers to each other are grouped to be fed together, a considerable amount of fighting and disturbance occurs for some time-often long enough to affect the ultimate financial outcome-unless the most timid are removed or effectively protected.

For that reason it is advisable to obtain each lot from one herd and to have all individuals of about the same body weight. When practicable, they should be dehorned, for it greatly simplifies the process of settling down. Under even the best conditions of yard feeding cattle which have been taken off adequate grazing will lose weight for a week or two.

The tropical feeder of individual cattle takes what comes his way: the barren cow, the injured bullock, the sterile bull, and makes what he can of them by feeding them on the by-products of his crops and the cereal offal from his household, by husbanding their energy and giving them the food which is just better than what they would otherwise have.

Feeding

The theoretical requirements of fattening cattle in the tropics have not been ascertained. Morrison's standards apply to temperate conditions, but they may be taken as an indication of the possible maximum amount of nutrients a thriving young beast can deal with under favourable tropical conditions.

Table 7.8: Nutritional requirements of fattening cattle To gain 0.75 to 1 lb (0.34-0.45 kg.) daily

Live-weight lb. (kg.)	*Dry matter lb. (kg.)*	*Digestible protein lb. (kg.)*	*Total digestible nutrients lb. (kg.)*
300 (136)	7.0-8.3 (3.2-3.76)	0.52-0.58 (0.24-0.26)	3.9-4.6 (1.77-2.09)
400 (181)	8.7-10.3 (3.95-4.67)	0.63-0.70 (0.29-0.32)	4.8-5.7 (2.18-2.59)
600 (272)	11.7-13.9 (5.31-6.31)	0.79-0.88 (0.36-0.40)	6.5-7.7 (2.95-3.49)
Fattening cattle			
800 (363)	19.6-22.2 (8.89-10.07)	1.46-1.62 (0.66-0-73)	14.1-15.9 (6.40-7.21)
1,000(454)	22.0-25.0 (9.98-11.34)	1.65-1.85 (0.75-0.84)	16.5-18.5 (7.48-8.39)

In practice, the feeder's aim must be to see that his stock are fed to appetite and that the appetite is stimulated and maintained as far as it can be by natural means, that is, by offering food which is readily acceptable. Many grass-fed beasts will, at first, refuse to eat food to which they are not accustomed.

Most oil cakes are not immediately acceptable; some, such as palm-kernel cake, are not eaten in quantity for a considerable time; but, generally, cereals and their by-products are readily taken. Some individuals persistently refuse to eat anything but grass, hay or straw, and are therefore unsuitable for stall feeding.

Except for the occasions when an animal is fattened for ritual or ceremonial slaughter, stall feeding is practised mostly by those who have a crop to dispose of or are able to obtain feed at an economical price, those, for instance situated near a cotton-ginning mill or an oil-extraction plant.

Their interest lies in their ability to supplement the main feed with a view to balancing it, and in this they may be guided by the figures given above. It will be noted that the amount of protein required by the fattening animal is very small-a mature beast needs practically none beyond its maintenance requirementsand so protein-rich food should be given only when it can be obtained so cheaply that its use for energy production is economical, or when it is desired to increase the palatability of other less tempting feeds.

ESTIMATION OF BODY AND CARCASE WEIGHT

Very experienced cattle dealers and butchers can judge the probable carcase weight of live cattle with astonishing accuracy by sight but such ability is not at the command of the average stock-owner, and unless he has some guidance his estimate of live-weight is a guess which may be very wide of the mark.

Certain body measurements when properly applied give an indication of live-weight which is remarkably accurate. The best-known and most widely used method is that of Shaeffer, expressed in the formula:

$$\text{Live} - \text{weight in pounds} = \frac{L \times G^2}{300}$$

when *L* is the length in inches from the point of the shoulder to the pin bone, and *G* is the girth in inches. We have found that estimates made on this basis of the weight of zebus between 300 and 900 pounds (136-408 kg.) live-weight have seldom been wrong by more than 10 per cent and usually are correct to within five per cent, the weight being more often under- than over-estimated.

It is not always easy to make accurate measurements of either the girth or the length of even a quiet animal, and in the field a difference of an inch or two between the measurements of the same animal by different people is the rule rather than the exception.

Aggarwalla recommended that the formula be modified as follows for use in India:

$$\text{Live} - \text{weight in seers} = \frac{G \times L}{Y}$$

where Y equals

9.0 if the girth is less than 65 inches,

8.5 if the girth is between 65 and 80 inches,

8.0 if the girth is over 80 inches,

and a seer equals 2.03 pounds (0.93 kg.).

Bennett (1951), working at Utah Agricultural Experimental Station, derived the following formula applicable to steers with a heart girth of 63 to 80 inches:

Live-weight in pounds = 1.04 (27.5758 × heart girth in inches) – 1049-67.

When the live-weight is known, an estimate of the dressed carcase weight can be made according to the 'condition' of the live animal. The dressed carcase from cattle in really fat condition ranges from 62 to about 65 per cent of the live body weight; that from cattle of the best indigenous breeds, in good condition, is about 54 to 56; and that from cattle in poor condition is 50 per cent or less. The majority of African cattle probably have a dressed out percentage of 47 to 50.

Bennett derived the following formula for calculating dressing percentage:

Dressed carcase weight = 44.08 – (0.0029 × weight in pounds) – (0.115 × height in centimetres) (0.2658 × heart girth in centimetres) – (0.0801 × paunch girth in centimetres).

DENTITION OF CATTLE AS AN INDICATION OF THEIR AGE

The eruption of each pair of teeth takes place approximately at the same time in life and thus an indication of an animal's age may be obtained by an examination of its teeth. As far as cattle are concerned, the indication is a very rough one because differences in age of as much as 16 months may be found in cattle with teeth at the same stage of development.

Differences of that degree are unusual, but variations up to six months are to be expected. The front teeth of cattle are easily examined and it is therefore to them that attention is generally confined, the stage of development of the molar teeth being noted only when further or confirmatory evidence is required.

On each side of the lower jaw there are four incisors, or front teeth, and six molar teeth in the adult bovine. In the upper jaw there are the same number of molars but no incisors.

Many factors influence dentition, the chief of which are breed and the standard of nutrition on which the animal has been reared. There is a very marked difference between the ages at which the teeth are erupted in cattle of the early-maturing breeds of temperate origin and the late-maturing breeds of the tropics. There is also a difference, but a less marked one, between breeds of tropical cattle and groups within the breeds which have been raised on different kinds and amounts of food. In addition there are individual variations at least as great as the average between groups, which cannot be specifically accounted for.

Table 7.9: Age at eruption of permanent teeth of cattle.

Teeth	*Age in months*			
	Indian farm cattle	*African farm cattle*	*U.S.A. ranch cattle*	*English pure-bred cattle*
First incisors	24-30	28	24	21
Second "	36	34	30-36	27
Third "	48	41	42	33
Fourth "	54-60	49	54-60	39
First molars	24		24	24
Second "	24		24	24
Third "	36		33	33
Fourth "	6		6	6
Fifth "	18		12-15	12-15
Sixth "	24		21	21

Lall (1948) collected information submitted by government-farms staff who examined farm cattle grouped according to their breed and recorded age. Because of the method used and because on some farms births were recorded only according to the quarter of the year in which they occurred no exactness can be claimed for the data collected.

The information he received referred to most of the important Indian breeds and he considers that the figures he gives, which are shown in Table elsewhere in this chapter, approximate the average closely enough to be of practical use.

In tropical Africa recorded observations are few and refer to but a small number of cattle. The usual individual and breed differences were noted but the average of the groups recorded by two observers Joubert and Kikule are very similar. The average ages at eruption of the permanent incisors of cattle of African breeds appear to be those shown in Table elsewhere in this chapter.

Observation of the ages at eruption of the permanent incisors of 78 White Fulani reared on a high plane of nutrition in Nigeria indicated that the average quite closely approximated those of mixed East African

Shorthorn zebu and Ankole at Entebbe, Uganda, but there were several individuals far removed from the average.

The corresponding figures for 'highly bred stock' in England and for ranch cattle in U.S.A. are also shown in Table elsewhere in this chapter.

When examining the teeth of cattle there may be some difficulty in deciding if the four pairs of front teeth are old temporary or permanent ones, but if there are six molars at each side of the jaw it confirms that the front teeth are permanent.

WORK OXEN

Machines are very quickly removing the burden of road transport work from animals but, as yet, machinery has not displaced animal draught power on the farm in tropical areas to any but an insignificant extent. The ox and, to a relatively small degree, the camel and the horse are still the chief sources of power in those areas where man himself is not saddled with the work.

The introduction of mechanization to the farm is delayed mainly by the smallness of the land unit held by the average farmer, the capital cost of machinery and the constant output of cash for fuel and upkeep, the difficulty or inability of switching machinery from one kind of work to another, the unsuitability of much of the mechanical equipment at present available, and the ignorance of the tropical cultivator concerning machinery.

As the amount of transport work demanded of animals decreases, the number of camels, donkeys and horses, especially the latter, correspondingly lessens, but the need for the ox and buffalo very largely remains.

Work-cattle maintain their popularity because of the ease with which they can be fed and maintained, the ability of the indigenous breeds to adapt themselves to the severest tropical climatic conditions, their hardiness, especially as shown in the relative freedom from, or power to withstand, digestive and lung troubles, in marked contrast to the horse and camel, the amount of work they can return for a given amount of food and the variety of work to which they can be applied.

As the use of the ox for long distance and heavy transport lessens, a change has to some extent taken place in the type of cattle required by the farmer; for instance, the large, long-legged Africander trek-ox is losing favour to the shorter-legged, smaller type of the same breed, as its use becomes more confined to domestic purposes. Similarly when the nomad settles and takes to cultivation, he forsakes the long-legged, foraging type for the more compact general-purpose breeds. The change is not made because work-oxen with small legs are better than those with long ones, but because the short-legged beasts are associated with breeds which mature quicker, fatten more easily and milk better, and as farm cattle find their food near at hand, the ability to forage is not of such importance.

The long-legged type of cattle is more esteemed in semi-arid areas where the soil is light and where work can be done at a quick pace; moreover, in those regions there is a greater intensity of radiated heat from the ground and the long legs lessen its effect upon the body.

On the other hand it might be thought that for such work as paddy cultivation on heavy irrigated land, length of limb would be an asset, but that does not appear to be so because in those situations the smaller breeds of buffaloes and such cattle as the Madras Ongoles are mostly used ; it must be admitted, however, that that may be a matter of necessity more than of choice.

There are good work-cattle of all shapes and colours, but the majority of good workers do conform to certain physical specifications and, when choosing draught oxen, if one cannot have the advantage of a test at actual work, one can, with a considerable amount of confidence, guess the good from the bad according to whether the animal has or has not the desired conformation.

The greatest source of error lies in the inability to judge the temperament of the animal without a prolonged test under the yoke. An unsuitable temperament will belie the promise of even the best conformation, while a suitable one will minimize, if it does not entirely overcome, the drawback inherent in physical defects.

Temperament is not to be judged on short acquaintance: an exhibition of nervousness or fierceness, for instance, which at first appears to be sufficient evidence of an undesirable character may actually denote the spirit or courage which all good workers require. In fact, it has been our experience with all species of work-animals that it is often those which initially are the most difficult to handle which eventually prove to be the best.

No animal which lacks sound, hard feet, and clean limbs with free action will give unqualified satisfaction, and these physical attributes should be the first to be sought for in stock selected to be trained for work. A powerful shoulder and a muscular, short neck with a well-developed hump make the fitting of the yoke easier and facilitates its carriage. Some of these features are associated with the mature bull rather than with the bullock or cow: the development of the hump is in particular linked to full masculine activity as is well demonstrated by its size in the progeny of zebu cattle crossed with European ones where the hump is obvious in the bull but non-existent in the cows and early-castrated males.

It is for this reason that the breeders of work cattle delay castration until the animal is fully grown or, at least, well grown, and by doing so come into conflict with the usual policy of government livestock departments which encourage the early castration of all bull calves not selected for breeding, with the object of preventing indiscriminate mating: early vasectomy would appear to be the way to solve the problem.

Bullocks generally make the most satisfactory draught animals: the bull is often unmanageable and the cow lacks strength. There are no other reasons why the cow should not work and, in fact, on light land and when properly fed, the dry cow gives quite satisfactory service in the plough; even a cow in milk may be used for moderate work for a few hours each day without greatly impairing her milk yield.

It is the bullock, however, which is the best suited both in regard to temperament and strength. Selected stock should be handled from the earliest age and may be put to light work when their bones are well grown but should not be put to heavy work until the bones are strongly knit: the completion of the process marks the full attainment of body size.

The age at which cattle are fit to start work thus depends upon how they have been treated during their growth, the quality and quantity of their feed, and their freedom from disease being the most important items in the process.

In cattle of breeds reared under hard conditions, such, for instance, as the Tharparkers, handling commences only at four years old and they are not put to full work until the age of five or even six years is reached. On the other hand, faster growing cattle, such as the Ongole, are put into training at two years and are in full work by three-and-ahalf years old if they have been reared under settled farming conditions.

The more they have been handled since birth, the easier they are to control and train, but it is astonishing to find how quickly even the most fractious cattle come to hand-with persistent, quiet, firm treatment, a few weeks is sufficient. Their training during that time must be directed to removing apprehension and then gradually to inducing them to answer direction of voice and hand.

After that has been achieved, too much must not be asked of them in the way of work or effort, for that leads to loss of self-confidence which results in refusal to try to do what is wanted of them, or in rebellion which has the same effect but is expressed differently: in the former case, the animal, on the slightest pretext, sinks resignedly to the ground and will not rise, and in the latter, aggressive refusal replaces docile compliance.

Head control is generally effected through a strong cotton rope, between five- and seven-sixteenths of an inch (0.8-1.1 cm.) in diameter, passing through a hole punched in the middle cartilage of the nose close behind the muffle, the ends passing up each side of the face and round the horns where they are spliced so that one completes the loop and the other is elongated to a rein of whatever length is desired.

The loop of the rope should be loose enough to allow of easy play but no looser. The device should not be fitted until the animal is well grown.

In placid breeds of cattle, a rope around the horn may be sufficient, or a halter like those used for horses may be worn. A rope halter can be constructed as follows : Tie a small loop at the end of a rope and a similar one about four inches (10.2 cm.) further along. Both loops should be just large enough to admit the free passage of the rope.

Now pass the free end of the rope through the first and second loops successively. When adjusting, another knot may be added at each loop to prevent slipping. The halter is particularly useful for the control of immature stock as it is persuasive more than compelling and excites little resentment.

The best drivers repeat a verbal order as they apply the control and gradually substitute the voice for the hand.

The earliest exercises should consist at first in leading and then in driving, at first singly and then as a pair. As soon as the animal learns to answer the control readily, it should be yoked with a steady, trained bullock and the instruction in the double yoke, without any load whatever, should continue until the young animal is quite familiar with the proceedings and has learned what is expected of him.

The yoke should then be attached to a very light load which should be increased gradually until it requires an appreciable amount of effort to move it. The training should be regular and not sporadic: it is better to do it for a short time twice or thrice a day than for a prolonged period.

The most serviceable yoke consists of a smooth, round pole some 56 inches (1.42 metres) in length with a diameter of at least three-and-a-half inches (8.9 cm.). It is fitted at the centre with a bolt or swivel to which the shaft can be attached, or with a projection to which it can be tied. Sixteen inches (40.6 cm.) on either side of this, a rod of very tough, smooth wood, one inch (2.5 cm.) in diameter is fitted to penetrate the yoke pole at right angles and to extend below it to a distance of one foot (30.5 cm.). Ten inches (25.4 cm.) beyond each of these, a flat-headed rod of iron, about

five-eighths of an inch (1.6 cm.) thick, or one somewhat thicker of very tough wood, is inserted parallel to the first ones, through the yoke pole in a manner which allows the rod to be moved up and down. These shoulder or neck rods come down at each side of the neck when the yoke is carried and keep it in position, maintaining the proper distance between the cattle.

The bearing surface, between the neck rods, is slightly bevelled so as to form a shallow arch which adapts itself more closely to the neck and increases its area of contact the edges are rounded off.

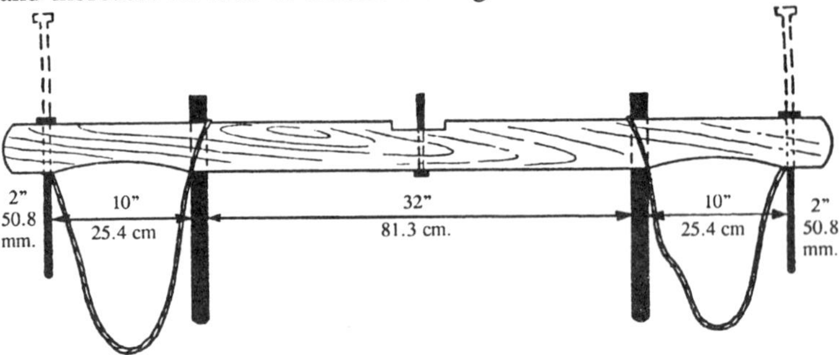

Figure 7.10: Double neck yoke.

The shaft of the yoke is made of smooth, tough timber stout enough to bear whatever load it is meant to take and to minimize the chance of damage through bruising or cutting which a narrow or angular structure is more likely to inflict when bumped against.

In cart work it is important to have the vehicle so balanced that when empty and on the level its weight lies forward on the necks of the oxen just enough to avoid the possibility of it being upturned: a thong slung under the necks from the head of one shoulder rod to the other, or a light bar running parallel to the yoke and attached to it through the lower end of the shoulder rod, prevents such an accident.

Many variations of this simple yoke have been designed with a view to increasing its efficiency, by making it fit comfortably over a larger area of the neck when under pressure, by lightening its weight, increasing its strength, and by removing projections which may cause bruising or impede easy handling. In practice, none, save the last, of these refinements seems to confer any great benefit.

It is asserted that if the bearing surface of the yoke is appropriately arched so that some of the weight is taken from the top of the neck on to the muscular part of either side, it will not only prevent pressure galls but will add to the general comfort of the animal. Such a yoke, however, does not conform to the thickening of the muscular mass of the neck when the head is lowered and the neck braced to take extra weight, so pinching and friction of the skin is then encouraged rather than prevented.

Vaugh has made a detailed study of seventeen modifications of the yoke described above and he came to the conclusion that although some are definitely better than others, he could find no clearly defined reason for the superiority except that yokes which restrict side movement of the bullocks are defective.

Less commonly the ox is worked by itself and not as one of a pair. For cart work this necessitates the use of double shafts which are supported on the neck of the bullock by a substantial arch of wood or galvanized tubular metal of sufficient bore to afford adequate bearing surface. For field work, traces must be substituted for the shafts and the yoke must be kept in position by a thong round the neck and by side strings to a surcingle or girth strap. The contrivance is insecure and troublesome to use.

Leather harness either in the form of a breast strap or a shoulder collar, as used with horses, has been employed but, presumably because the shoulder muscles of bullocks are not so well developed as those of horses and thus neither protect the joint nor cushion the skin, it appears to cause discomfort and discourages traction.

A head yoke is more satisfactory. Details of the construction of a simple and apparently effective one designed by Read at Hissar, India, will be found in Appendix III. This design is described in full because the contrivance is not well known and it is thought that by its use a single ox may be used where a pair is often employed merely because a convenient single yoke is not at hand. Animals adjust themselves to the head yoke with little training.

Yoke galls are the chief cause of wastage in work-oxen. They are particularly liable to occur during the first weeks of work before the skin has become accustomed and hardened to the almost inevitable friction, which is exaggerated by the unsteady traction exerted by the inexperienced animal. Deeper injuries arise from bruising or from excessive pressure exerted by the weight of the load for prolonged periods when the underlying tissue is deprived of its full blood supply and is consequently liable to destruction.

A yoke with a broad, smooth bearing surface allows the weight to be distributed over a larger area of skin and is less likely to cause a pressure gall than is a narrow one. Galls can be avoided or mitigated by ensuring that the periods of work at first are short, that the load is removed from the neck during periods of rest and halts of any appreciable length of time, that the load is properly balanced and that no unnecessary weight is thrown on the neck, that both the skin and the yoke are kept clean, and that the skin is greased when working in the rain.

A pair of good oxen should be able to do all the draught work required on a holding of 15 acres (6.1 hectares): they commonly serve an area of 20 acres (8.1 hectares) of cultivated land.

Cattle, like other animals when left to their own devices, have regular periods of increased activity and of rest throughout the twenty-four hours. If these are to be harmonized with those of man in hot weather, work on the farm should be confined to the early hours of the day and to the evening hours as the heat of the day declines.

A seven-hour day with an iron plough is as much as can be expected if it is to be maintained over any prolonged period and it is best performed between six and 11 a.m. and four to six p.m. With a primitive wooden plough an average of nine hours can be worked.

For long periods of the year the cattle will be idle, while at other times, weather permitting, a 10-hour day and more can be accomplished. Working such hours, it is necessary that the oxen should be largely stall-fed unless good grazing is available at night. If the cattle are to find their own fodder, they must be given at least seven hours daily to do so and no more than seven hours' work should be expected of them over periods of any length.

If they are on transport work during hot weather it should, whenever possible, be done during the night so that the heat of the day may be avoided: it is, moreover, the normal period of greatest activity of tropical cattle on free range. Four hours should be done in the early evening followed by a rest of four hours, and the work should be completed for the day by the time the sun is well up.

Joshi and Phillips report that a pair of Hariana cattle can pull one ton (1.02 m. tons) for 20 miles (32.2 km.) at two miles (3.2 km.) an hour: that a pair of Bhagnari cattle are expected to move one-and-a-half tons (1.53 m. tons) on a primitive cart on a dirt track for 20 to 25 miles (32.2-40.2 km.) each day at the rate of two to two-and-a-half miles (3.2-4.0 km.) per hour and that Hallikar cattle have done up to 40 miles (64.4 km.) on rough roads pulling an average load.

On the other hand, Hattersley (1951) considers that a work day of five hours in a ridging plough is sufficient in the Sudan for a pair of cattle, 'even in extreme heat', if properly fed, but he considers that an occasional two to three extra hours may be demanded.

When draught oxen were used in the British army in South Africa, they covered 15 to 20 miles (24.1 to 32.2 km.) day after day at the rate of two to two-and-a-half miles (3.2-4.0 km.) per hour and found their own food, when they were well managed: in India, under favourable conditions, they travelled up to 24 miles (38.6 km.) a day.

Work in the plough can seldom if ever be kept as steady or at as good a pace as draught work on the road. How much can be done in a day on the farm depends primarily upon the condition of the cattle; the nature of the soil to be cultivated; the kind of vehicle to be pulled, e.g. plough, harrow or cart; the weather; the method of feeding and watering; the competence of the driver and on many other lesser factors.

With an eight-inch (20.3 cm.) furrow, it should, in theory, be possible to plough one acre (0.40 hectares) in six-and-ahalf hours if the pace of two miles (3.2 km.) per hour can be maintained, but because of the many inevitable distractions connected with such work, twice that time may be needed. In fact, the conditions are so variable that no hard-and-fast rule can be imposed and it will be found that the time and loads that are general to the district are those that it is safe to follow until experience teaches how they may be changed.

An over-worked animal loses weight. If work is carried out at the expense of the animal's body weight it is only a question of how soon the work will come to an end: the position may be righted, presuming the animal is healthy, by correcting faults in management or in working conditions, apart from lessening the work hours.

The difficulty usually experienced is in detecting the loss of weight soon enough to correct the position before it is serious. It has to be remembered that food of animal origin is one of the most expensive, and that an animal losing condition through work is living on the tissues of its own body: moreover, it cannot be expected to give full work while it is rebuilding its body, and thus early detection and prevention of loss of body weight makes the difference between economic and uneconomic labour.

In practice, if there is any doubt about a possible loss of condition, as judged by eye, the loss should be assumed and measures taken to right it. It is seldom that facilities for weighing cattle are available outside a research station, but if measurement by weighing can be done, it must be remembered that a sporadic loss of twenty or thirty pounds (9.1 or 13.6 kg.) in weight may indicate nothing more than that the animal has defecated and urinated before being weighed on one occasion and, on another, has had a drink or a feed beforehand.

Only a sustained, consistent loss of weight over a period is significant, and that loss is generally apparent to an experienced eye. Moreover, a loss of body fat, the material first used to substitute a shortage of food, is associated with an almost immediate deterioration in general appearance as shown by a lack of lustre in the coat and a want of zest in activity.

Inexperienced animals expend more energy in doing a given amount of work than they do when they have become accustomed to it, and they need particular attention in their period of training and in the early days of work. Each work-ox has a rate of movement best suited to its gait and if it is hurried out of it the expense in wear and tear is excessive: an average of two to two-and-a-half miles (3.2-4.0 km.) per hour is normal.

A pair of oxen should be matched in temperament as well as in physical characters. A good ox-driver takes note of the idiosyncrasies of his charge and allows for them: he provides shade when resting, an adequate supply of drinking water, the odd mouthful of food during pauses in work, and he maintains an unhurried and imperturbable relationship. Apart from food and health, it is his treatment of his oxen upon which their condition depends more than anything else. When well managed they should work satisfactorily until they are about 17 years old.

FEEDING

Chiefly because of the unpredictable variation of factors which determine the exact amount of work an animal is doing, and therefore the energy expended while doing it, when not under laboratory control, no standard of nutritional requirement has been set in terms of calories or other exact units. In fact, it is not definitely decided how much, if any, increase of digestible protein is directly required for the output of work, but experiments indicate that probably none or very little is needed.

Lander calculated tentatively, on a basis which is open to criticism, that 0.327 pounds (0.148 kg.) of T.D.N. with 0.0297 pounds (0.0132 kg.) of D.P. are needed by a 1,000-pound (453.6 kg.) bullock for

Table 7.10: The maintenance requirements of work oxen.

Body weight lb. (kg.)	*Total digestible nutrient* lb. (kg.)	*Digestible protein* lb. (kg.)
500(226.8)	4.52(2.05)	0.248 (0.112)
600(272.2)	5.16(2.34)	0.284 (0.129)
700(317.5)	5.78(2.62)	0.317 (0.144)
800(362.9)	6.37(2.78)	0.350 (0.159)
900(408.2)	6.94(3.15)	0.381 (0.173)
1,000 (453.6)	7.49(3.40)	0.412 (0.187)

each hour of work in the Punjab, but the figures should be taken as only a rough indication of average requirements until they have been confirmed. When they are added to the maintenance requirements, the total amount needed for the average work-ox are indicated. The maintenance requirements, according to Lander, are shown in table elsewhere in this chapter.

Lander also states that in India the following rations are suitable for 'three typical classes of bullocks.

1. Bullocks weighing approximately 800 pounds (362.9 kg.) and working a six-hour day in Southern India.

   Rice straw 10 lb. (4.5 kg.)
   Green sorghum 20 lb. (9.1 kg.)
   Mustard *(Brassica juncea)* 2 lb. (0.9 kg.)

   or

   Sorghum straw 15 lb. (6.8 kg.)
   Rape *(Brassica napus)* 1 lb. (0.45 kg.)

2. Bullocks approximately 1,000 pounds (453.6 kg.) live-weight, ploughing for six hours in the hot season.

   Wheat bhusa (chopped wheat straw) 5 lb. (2.3 kg.)
   Jowar *(Andropogon sorghum vulgare)*
   and Guara *(Cyamopsispsoraloides)* 25 lb. (11.3 kg.)
   Gram *(Phaseolus spp.)* 4 lb. (1.8 kg.)

   This ration is equivalent to 16.42 pounds (7.45 kg.) of D.M. containing 9.45 pounds (4.29 kg.) of T.D.N. and 0.59 pounds (0.27 kg.) of D.P.

3. In the Punjab where are the best cattle in India, the village workoxen get the following types of ration:

   Wheat bhusa 18 lb. (8 2 kg.)

Rape cake	2 lb. (0.91 kg.)
or	
Jowar stalks	18 lb. (8.2 kg.)
Guara grain	2 lb. (0.91 kg.)

When grazing is at its best, and usually that is the season when most draught work is required on the farm, a bullock can maintain its condition while doing seven hours' medium work each day, provided it is allowed at least seven hours in which to graze, and during the remaining hours of the 24 it has sheltered rest.

In other words, an 800-pound (362.9 kg.) bullock will eat 16 pounds (7.3 kg.) of D.M. which may be had from 53 pounds (24 kg.) of mixed pasture grass which will contain about 10.5 pounds (4.76 kg.) of T.D.N. and 1.76 pounds (0.80 kg.) of D.P.-an ample supply according to Lander's calculations and to our experience.

In the dry season, when there is little farm work to be done, and grass is scarce and innutritious, an idle bullock may be just able to maintain its condition, but in difficult circumstances the energy in collecting the fodder will outweigh the amount obtained from it and it must be supplemented.

Such material as cereal offal, *e.g.* the bran of rice, sorghum, etc., or even good quality broken straw is suitable if hay or silage is not available.

Hartley and Ross (1938) considered that in the dry weather in arid Northern Nigeria, 7 pounds (3.2 kg.) of fodder, of which half should be legume straw, is required to supplement normal grazing to keep an idle bullock in condition, and that if it was working during that time of the year, it needed in addition one pound (0.45 kg.) of *dhusa* (sorghum bran, containing about 0.744 pound (0.337 kg.) T.D.N. and 0.057 pound (0.026 kg.) D.P.) for each hour's work.

They also recommend that three or four weeks before the ploughing season begins, oxen should be fed three pounds (1.4 kg.) *dhusa* or three pounds (1.4 kg.) of fodder and one pound (0.45 kg.) of *dhusa* in addition to the maintenance ration, to get them into condition.

Commonly, little attention is paid to the idle work-bullock, and a loss of weight is considered of small account, in the belief that if full rations are given when full work is required the response will be immediate and satisfactory. That is, however, not the case: if an animal in poor condition is put into full work no amount of food will prevent it losing further weight initially.

On the other hand, a fat animal is not usually a good worker and no benefit is to be derived from giving an idle ox more than will keep it in thrifty condition.

The work-ox should never be kept short of drinking water. In the hot weather it should be offered at least three times a day but, in other seasons, if it is given in the morning and evening, that will suffice.

8

BUFFALO

The domestic buffalo or water buffalo belongs to the *Bovidae* family and from time immemorial has been a domestic animal in India, Malaya and Egypt. It occupies an important place amongst the domestic animals of the tropics as a provider of dairy produce, beef and draught power. Its greatest asset as a domestic animal is its ability to subsist on the coarsest fodder and to convert it most efficiently into animal produce.

The water buffalo is found in a more or less domesticated state throughout the northern tropical and sub-tropical zones in Asia as well as in the Philippines and Trinidad and in all countries of the Mediterranean basin except France.

In the southern tropical zone it is found in Indonesia and it has been introduced to Melville Island near the northern coast of Australia. It is also found wild in the Northern Territories of Australia but the original animals were domesticated.

In recent years, the buffalo commonly known as an 'Asian Animal' has attracted global concern. The buffalo is the dairy, draught and meat animal of Asia. The Indian subcontinent is the home tract of the world's dairy buffaloes. To-day in India, the water buffalo is recognized as her milk machine.

It accounts for more than half of India's total milk production, although it constitutes only one third of the total milch population. Nevertheless, its potential for milk production remains only partially tapped, as almost 80 to 85 per cent buffaloes are non-descript, yielding on an average no more than 1.5 litres of milk per day in contrast to an average of 7 to 8 litres of daily milk yield obtained from well bred milch breeds.

The domesticated buffaloes may be classified into two main categories, viz., (1) the swamp buffalo (chromosomes, 2n = 48) and (2) the river buffalo (chromosomes, 2n = 50). They belong to the same species, *Bubalus bubalis* but have very different habits.

Some Common Terms in Relation to Buffalo

Details	*Expressions*
Species called as	Bovine of Butaline
Group of animals	Herd
Adult male	Buffalo bull
Adult female	She-buffalo or buffalo cow
Young male	Buffalo bull calf
Young female	Buffalo heifer calf
New-born one	Buffalo calf
Castrated male	Buffalo Bullock
Castrated female	Spayed
Female with its offspring	Calf at foot
Act of parturition	Calving
Act of mating	Serving
Sound produced	Bellowing
Pregnancy	Gestation

As the name implies, the water buffalo, whether of the river or the swamp type, has an inherent predilection for water and loves wallowing in water or mud pools. The river buffaloes, as a rule, shows preference for clean running water, whereas the swamp buffalo likes to wallow in mudholes, swamps and stagnant pools. Like cattle buffaloes are good swimmer.

The river and swamp buffaloes vary slightly in body structure, size and colour. A swamp buffalo is recognised by its short stocky body, short face with wide muzzle and short thin legs. These buffaloes are usually dark grey in colour, although variations from black to albinoids are found among the population. Swamp buffaloes vary in size from 300 to 600 kilograms. They are primarily used as draught animals, but they do provide small quantities of milk for a farmer's family.

BREEDS OF INDIAN BUFFALOES

Only in India are there found well-defined tropical breeds with standard qualities and, even there, well-bred individuals are far out numbered by nondescript animals. In other countries, although distinct types are to be found, no type has the specific physical characters which mark it from other types so that it can be unquestionably recognized. Good, well-defined, milk breeds are found only in the Punjab, Rajasthan, and the Gujarat district of Bombay, but in Central and South India there is a noted draught breed.

These animals belong to a non-descript class and these vary greatly in size, weight and general features known as *deshi*. However, on the basis of regions the well defined buffalo breeds are as follows:

Table 8.1

Murrah Group	*Gujarat Group*	*Uttar Pradesh Group*	*Central India Group*	*South India Group*
Murrah	Surti	Bhadawari	Nagpuri	Toda
Nili-Ravi	Jaffarbadi	Tarai	Pandhepuri	South Kanara
Kundi	Mehsana		Manda	
Godavari			Jerangi	
			Kalhandi	
			Sambalpur	

Murrah

The *Murrah* is the most important Indian breed of buffalo, and is the most efficient producer of milk, not only in India but probably in the world. Bulls of this breed are used extensively for up-grading inferior stock. She-buffaloes are used in most of the important cities for the supply of milk and ghee.

The home of the breed is mainly in Punjab and Delhi, but animals are bred pure in the United Provinces, Rajasthan and other places. Rohtak in Punjab is a well-known market from where thousands of high yielders are exported.

The Murrah has a deep massive frame with short, broad back and a comparatively light neck and head. It has short, characteristics tightly curled horns, well-developed udder and long tail with a white switch reaching to the fetlock.

Other features are: short massive limbs with good bone, broad hoofs and drooping quarters. The colour is usually black, but fawn-grey colouring of the hair and white markings on the face, legs or tail are not uncommon. 'Wall' eyes are also not unusual.

The skin is soft, smooth with scanty hair. The body weight of bulls amounts on the average to 550 kg that of the she-buffaloes to 450 kg. The height at withers on the average is 1.42 m for bulls and 1.32 m for she-buffaloes.

Production

The average milking capacity in established herds is about 1400 to 2000 kg with a butter fat content of 7 per cent of milk in a lactation of 9 to 10 months.

Nili/Ravi

The *Nili* and *Ravi* are two types of buffaloes found in the valley of the rivers Sutlej and Ravi in the Districts of Montgomery and Ferozepur. There is no essential difference between the two types although for a long time they were treated as different breeds, but closer study indicated that it was a distinction without a difference, and now they are officially treated as one breed.

This is considered to be one of the best breeds of buffaloes in India, next only to the Murrah. Like the Murrah large numbers are exported annually from the breeding areas for milk production in cities. The average milk yield is practically the same as that of the Murrah.

These animals have a medium-sized, deep frame with an elongated, coarse and heavy head, bulging at the top, depressed between the eyes and ending in a fine muzzle. Horns are small with a high coil and the neck is long, thin and fine. It is the face and forehead that distinguish this breed mainly from the Murrah.

The typical Nili/Ravi cow has a well-developed udder. The tail is long, almost touching the ground.

The colour is usually black, but brown is not uncommon. Pink markings are sometimes seen on the udder and brisket. White markings on the forehead, face, muzzle and legs, white switch and wall, eyes, which are also seen, are much liked by breeders. The average live-weight and measurements of mature animals are as follows:

	Male	*Female*
Live-weight : pounds (kg.)	1,300 (589.7)	1,000 (453.6)
Height at withers : inches (cm.)	54.0 (137.2)	53.5 (135.9)
Length from point of shoulder to pin Bone : inches (cm.)	62.5 (158.8)	58.5 (148.6)

Surti

The *Surti* is a popular breed found in the 'Charottar' tract of Gujarat in Mumbai State between Mahi and Sabarmati rivers. The Surti is known to be an economic milk producer and the average lactation yield is reported to be 3,650 pounds (1,655.5 kg.) milk with 7.5 per cent fat.

Surti buffaloes are well-shaped animals of medium size. They are rather low on the legs and have a mild and placid disposition, but the general appearance is bright, with prominent, round, rather bulging eyes. The horns are of medium length and are sickle-shaped.

The colour is black or brown; good specimens have two white collars, one just round the jowl from ear to ear and a second lower down around the brisket akin to the breast plate worn by horses. No other breed of buffalo has as straight a back as this one.

The udder is well shaped and well developed with pinkish-coloured skin and medium-sized teats, placed squarely. The skin is fairly thick but pliable, soft and smooth, with scanty hair.

The average measurements of mature animals are as follows:

	Male	*Female*
Height at withers : inches (cm.)	51.5 (130.8)	49.0 (124.5)
Length from point of shoulder to pin Bone: inches (cm.)	56.3 (154.2)	54.5 (138.4)

Jaffarabadi

Jaffarabadi buffaloes are very massive animals and are seen in their purest from in the Gir forest of Kathiawar where they are bred in large numbers, almost solely for ghee production. They consume large quantities of fodder but the milk yield is high and the butterfat content exceptionally high.

Their noticeable feature is the very prominent forehead and heavy horns which are inclined to droop on each side of the neck and then turn up at the points but not in such a tight curl as in Murrah buffaloes. They are usually black in colour.

Mehsana

This breed is found in districts of Bombay State and neighbouring areas. It is considered to be an intermediate type between Surti and Murrah buffaloes but here is a considerable amount of variation from one district to another. The horns of these animals are usually curled after the manner of Murrah buffaloes but not so tightly.

They are either black or fawn-grey in colour with usually some white marking on the face, legs, or tip of the tail. The udder is usually well shaped with uniformly placed teats. Mehsana buffaloes are considered to be particularly valuable milch cattle because of their early maturity, persistence in milk production and regularity in breeding.

They are of medium size, economical to feed, and are used largely for milk and ghee production in Bombay state.

UTTAR PRADESH GROUP

Bhadawari

Bhadawari estate of Agra district and adjoining areas of Gwalior and Etawah. Animals are found scattered in the surroundings of Jamuna and Chambal rivers.

Medium size and wedge-shaped body. Comparatively small head bulging towards horms. Legs are short but stout. Hooves of the hind quarters are more backward than fore quarters. Colour is copper, hair scanty. Barrel is short but well developed.

Table 8.2

Trait	*Range of vale*
Age at first calving (months)	48.3—50.78
Lactation yield (lit.)	1111—1252
Lactation length (days)	276
Calving interval (days)	453.6
Dry period	156

Forehead is slightly broad and deep in the middle. Face is comparatively narrow with slightly marked nose. Eyes are prominent, active and bright. Udder is not so well developed but milk veins are fairly prominent. Teats are of medium size and uniform in length.

Average yield ranges from 2,000 to 2,070 kg. in a lactation period of 305 days with a high fat percentage. Males are used for draught purposes.

Tarai

The breed gets the name from the *Tarai* area of U.P., where it is mostly found between Tanakpur and Ramnagar. The breed is native to hilly area. They are frequently crossed with Murrah bulls.

It has a moderate body, slightly convex head with prominent nasal bones. Horns are long and flat with coils, bending backwards and upwards having pointed tips. The eyes are rather small but ears are long and coarse. Legs are short but strong. The tail is long, reaching below the hocks. The colour of the skin varies from black to brown. Sometimes there is a white blaze on the forehead. The switch of the tail is white.

The breed is poor regarding milk production, which may be as low as 2-3 kg daily. Males are efficient draught animals used for agricultural operations as powers including transport. Tarai breed is well adapted to the difficult environmental situations and nutritional conditions of the Tarai.

CENTRAL INDIA GROUP

Nagpuri

Synonym, Marathwada, Berari, Ellichpuri, Gaulani, Varad, Gauli.

As the name applies, these animals are mostly found in Nagpur including Wardha, Yeotmal, Akola, Buldana, Amrawati and Achalpur in Maharastra. The breed has got several strains differing in colour and other superficial traits.

The colour is usually black; occasionally white markings are observed on the face, legs and switch. In general the breed is known as lighter type than those of north, not so squat and have a short tail. The face is long and thin with a straight profile. The neck is also longer with heavy brisket. It has long horns flat-curved reaches towards back over the shoulders. Naval flap is short or almost absent.

Male body weight averages 525 kg. while female attains about 425 kg. The average height at withers in male 142 cm and that of cows 132 cm; heart girth is 210 and 205 cm. for male and female respectively.

The females are moderately good yielders, producing on an average 1,000 litres milk per lactation of 300 days with butter fat content between 7.0-8.5 per cent. The males are used for heavy draught but are slow.

Pandhepuri

Synonym Pandharpur, Dharwari

The breed is widely distributed in south Maharastra, parts of Andhra Pradesh and Karnataka States.

Animals are of medium size with long narrow face and very long, flat and usually twisted thin horns.

Males are hardy and well suited for drought purpose.

Manda

Synonym, Parlakimedi, Ganjam.

The animals are bred in the hills above Parlakimedi and Mandasa on the borders of Orissa and Andhra Pradesh. Basically the breed is reared in areas of thick forest usually on natural herbage and brought down to the plains for sale.

The general colour is brown or grey with yellowish tufts of hair on the knees and fetlocks, and the switch is yellowish white. The breed is a medium sized animal. The eyes are sharp with a broad red margin around the lids. The horns are broad and semi-circular extending backward and inward. The forehead is flat with short muzzle. Neck and forelegs are also short but stout with well developed chest. Milk yield is satisfactory. Males are hardy and like a bullock can work in the hot sun. It can draw a load of about a ton but at a slow pace.

Sambalpur

Synonyms, Kimedi, Gowdoo.

The home tract of this breed is controversial. Originally it was known to be habitat of Sambalpur area of Orissa, later on it has been suggested that the main habitat of this breed is around Bilaspur district of Madhya Pradesh wherefrom calves are brought by 'gowdoo' (Herdman) to Sambalpur area.

Animals are large and powerful having long, narrow barrel and prominent fore-head. Body and coat colour is generally black but it varies to brown and ash grey.

Males are very active and good for drought purposes which are affected by high atmosphere temperature. Females breed regularly and produce milk satisfactorily.

Kalhandi

Synonym, Peddakimedi.

The home tract of this breed has been referred to the eastern part of Andhra Pradesh and adjoining areas of Orissa.

A strong with broad horns and half curved running backward at the tip. Body colour is between grey and ash grey. It has well developed chest and strong fore-limbs. Eyes are prominent and large with narrow red margin around the lids. Tail is of medium length with white switch. Due to light colour it tolerates heat more than dark coloured buffaloes.

Animals are docile and hardy. They are used to carry loads in hilly areas as well as for ploughing in plains. Milk yield is satisfactory.

Jerangi

This breed of buffalo is widely distributed in Jerangi hills of Orissa and northern parts of Visakhapatnam and west of Ganjam in Andhra Pradesh. One of the dwarf breeds of buffalo and its height does not exceed four feet. Horns are conical and small, and run backward, body colour is black.

Not that much good in milk production but are useful animals for ploughing in water-logged paddy fields.

SOUTH INDIA GROUP

Toda

The name of the breed originates from the name of *Toda* tribes of Nilgiri hills of Tamil Nadu State who rear this breed.

These are large-sized animals having long barrel and strong build. Horns of these buffaloes are variable in shape but are usually set wide apart run outward, upward and then inward at the top. Face is short and wide. Hump and dewlap are absent, chest is broad and deep; legs are short and sturdy. Along the crest of the neck, hump area and back, there is a thick growth of hair like a mane which imparts a bison like appearance.

The animals are not always docile. All of sudden they may become furious and dangerous particularly to unknown persons. When necessary, they fight against even tigers and other wild animals in an organised team.

Females are good milkers, yields between 4.5-8.5 litres daily with an average of 7.0% fat and possesses typical well-flavoured taste.

BREEDING AND MANAGEMENT PROBLEMS

Buffaloes are Seasonal Breeders

In tropical countries like India, *she* buffaloes indicate seasonal sex periodicity in respect of oestrous and conception. As much as 62.8% animals show periodicity in October to February with the peak around December and this period coincides with higher conception rate. During the period from March to September, with comparatively longer days and higher intensity of sun, ovarian activity appears to be adversely affected.

Buffalo bulls are sexually least active during hot seasons. High environmental temperature upsets the normal physiological functions affecting spermatogenesis in males and ovarian activity in the females. 82% of services take place in November. Even under A.I. conditions, 65% of the inseminations of the year are being carried out during the months of September to February.

Thus 80% of buffalo calvings are recorded only in the months of July to December and minimum in hotter months. With judicious management practices such as heat detection, protection from thermal stress, adequate and balanced ration and optimum time of insemination, the calvings in buffaloes could be evenly spread round the year to a great extent.

Silent Heat

Silent heat refers to normal follicular development and ovulations without the behavioural signs of estrus.

In summer the extreme climatic stress coupled with increased day light reduces the incidences of buffaloes coming in regular heat. During this time other reproductive problems like repeat breeding and pregnancy losses are also increased. It is generally felt that most of the buffaloes do have sub-functional ovaries causing weak or silent oestrus (in 60% cases) and such incidences are even higher than the true anestrous, especially in summer.

The research workers are trying to find out the causes of these disorders of ovarian origin and are successfully trying to correlate them with the lack of management, nutritional status of the animal and hormonal (serum gonadotrophins and steroids) profiles.

However, if due to excessive environmental stress the CL in buffaloes does not regress or partially regresses (in non-pregnant condition), it will continue to secrete progesterone and will inhibit further release of LH, thus the buffaloes will go in a state of anestrus or subestrus conditions, so long as CL does not completely regress.

Corrective Measures

For animals which either do not manifest oestrus or are in a state of silent oestrus or sub-oestrus firstly need managemental attention rather than therapies.

Two or three examinations with an interval of a week between each, will reveal that most of them are cases of apparent anestrus. In such cases simple utero-ovarian massage and close detection of oestrus will elleviate these conditions in most of the cases.

In extreme summer condition these animals should be protected from direct solar radiation by maintaining them in shaded half walled sheds. If possible animals be given showers in addition to wallowing facilities particularly at mid-day during April, May and June.

Parading vasectomized bulls for heat detection can be gainfully employed at organised/large farms. Since in summer buffalo bulls soon lose libido, replacement by young high libido bulls may be thought of where it is applicable.

Painting of Lugol's Iodine

The external is swabbed thoroughly with Lugol's solution (B. Vet. Co, Iodum 5 g, pot, iodide 10 g, distilled water 100 ml) with the help of a metallic swab holder (46 cm.) and vaginal speculum- double application of this solution is sometimes recommended at week's interval. It is presumed that application of Lugol's Iodine on the cervix causes local irritation and bring about reflex stimulation of anterior pituitary for secretion of gonadotrophins and thus cyclicity starts.

Some workers have used estrogen and progesterone in anaestrus and sub estrus cows with encouraging success. The idea behind using small doses of these hormones are to act as a positive fed back at the hypothalamus level to release gonadotrophin releasing hormone (Gn-RH) and simultaneously sensitize the gonadotrophas in the anterior pituitary so that Gn-RH may act fully to stimulate more release of gonadotrophins.

It should be noted that the method can never be considered as a substitute for good management since use of hormonal therapy might also lead to inherent weakness in the progenies.

Use of prostaglandin ($PGF_2\alpha$) induces luteolysis resulting in sharp fall of serum progesterone level and thus helps to bring back normal oestrus cycle. Like hormonal therapy, prostaglandin therapy also can not be recommended as a substitute of good management.

Repeat Breeding

A repeat breeding in cow or buffalo is the one which has normal oestrus, oestrus cycles as well as reproductive tract and has been bred three or more times by a fertile bull or semen yet failed to conceive.

It is possible that pathology which cannot be detected by normal clinical methods may be present which include among others, failure of fertilization, an ovulation and delayed ovulation, tubal obstructions, early or latent embryonic mortalities, poor breeding and management techniques including genetic, nutritional and infectious factors.

Endometritis in many cases has been found to play major factor for causing repeat-breeding. Endometritis is caused by sporadic uterine infections by a varieties of microorganisms such as *C. pyogenes*. *Coliforms*, *Pseudomonas aeruginosa*, *Streptococci* etc., viral and fungal agents along with mycoplasmal agents may also cause endometritis in cattle and buffaloes. The condition follows mainly parturition, especially abnormal ones; abortion dystokia, retained placenta, genital prolapse, uterine inertia, traumatic lesions in the uterus, cervix and vagina.

Endometritis may also occur in cows and buffaloes after coilus or after AI under unhygienic conditions. Appropriate antibiotic treatment is advocated after conducting antibiotic sensitivity tests where repeat breeding is associated with infection of the tubular genitalia.

Daily examination of repeaters is necessary to correct the conditions arising from ovarian dysfunction. Anovular heats can be detected from the absence of palpable C.L. during dioestrus. Intramascular injection of 25 mg of progesterone during early heat may help inducing ovulation.

The golden rule, prevention is better than cure applies more appropriately to reproductive problems. The incidence can be kept under control by adopting methods such as: (1) improved feeding and management, (2) providing hygienic surroundings at the time of calving, (3) asceptic precautions at the time of inseminations, and (4) educating farmers on detection of heat and maintenance of breeding records.

Weaning

Separating calf from its mother at a very early age is a bit difficult in buffaloes due to high motherly instinct. Since the method is important to assess dam's milk production and to feed calf according to live weight a substitution technique could be followed.

A single calf is substituted for let-down of milks in a number of buffaloes and the real calves are weaned just after birth. It can be achieved by putting blinkers (materials used to shut off the side views of buffaloes) during parturition and subsequently soiliing the body of substituted calf with placental secretions. The dam starts licking the substituted calf after removal of blinkers.

Calf Mortality

Buffalo calf mortality rate under one month of age averages about 10% and varies from 3-30% in individual herds. Losses upto 50% have occurred in large dairy herds. A calf mortality rate of 20% can reduce net profit by 38%.

A number of factors considered responsible for this hazard, are discussed below:

Antibodies are not transferable from buffalo dam to her fetus through placental membranes and thus they are susceptible to various virulent diseases. In buffalo calves inspite of feeding colostrum, a low antibody titre exists. Immunoglobulin levels have been reported to be 29.73 mg/ml it day olds and the quantity increase to 35.66 mg/ml on the second day of life.

On the other hand in one estimate it has been noted that colostrum which is the only source of immunoglobulin for the buffalo calf, contains 68.75 mg of immuglobulin per ml on the first day, 23.75 mg/ml on the second day and 1.01 mg/ml on the fifth day of lactation.

Certain meteorological influences may have an effect on calf mortality rate. During the winter months, mortality may be associated with the effects of cold, wet and windy weather while during the summer it may be the hot, dry weather.

Most of the deaths occur during autumn and winter months before the age of 3 months. The causes of mortality in order of priority have been found (i) Pneumonia, (ii) Enteritis, (iii) Toxaemia/Septicemia, (iv) Worm infestation, (v) Bloat etc.

The cause of high mortality in male calves could also be due to neglecting tendency by the management particularly regarding feeding.

Some of the prescribed managemental practices including feeding practices may be followed for minimizing the mortality rate.

Housing

In commercial herds, calves after weaning may be kept in groups in large pens where individual feeding is advocated. In small herds calves after weaning should be kept in separate pens (24 sq. ft.) upto 3 months of age to avoid suckling instinct. The methods will eliminate the possible calf scour and parasitic infestation.

Feeding Colostrum

It is the milk secreted by the udder immediately after parturition and for the following 3-5 days. It contains 20% or more protein a little more fat, 10 to 100 times more vitamin A. three times more of vitamin D and may be tinged pink due to blood corpuscles. It acts as a natural purgative for the calf, cleaning from its intestines the accumulated faecal matter (meconium).

Table 8.3: Milk feeding schedule

Age (days)	*Colostrum (kg.)*	*Body weight (kg.)*	*Milk (liters)*
1–5	1 to 1.5	—	—
6–15		Upto 25 kg.	1.0
11–15		26–30	2.0
15 and above		31–40	2.5
		41–45	3.0
		46–50	3.5
		51–55	4.0
		56–60	4.5
		60 and above	5.0

Of much greater importance, it is through the medium of the colostrum, antibodies which protects against various bacteria and viruses are supplied to the new born calves. Colostrum coagulates at about 80^0 to 85^0 C and can not therefore be boiled.

The calves at birth after weaning should be fed colostrum within 2 hrs. and its feeding should continue for a period of five days at the rate of 1 to 1.5 kg per day. In case the dam does not give colostrum a substitute of equal nutritive value prepared from 2 eggs and an ounce of castor oil may be fed for builiding up resistance.

It may be necessary to inject dam's serum to such a calf for augmenting antibody titre in the body. The feed efficiency ratio of buffalo calves is as high as 1 kg. gain per 1.16 kg. dry matter. Milk feeding schedule is given in tabular form.

It is necessary to provide milk to weaned calves at body temperature preferably with supplements to make up the deficiencies of Fe, Cu, Mg, Zn and Mn. Green fodder containing upto 100 gram dry matter may be offered daily from 15 days of age onwards so that it stimulates early rumen development.

Feeding Antibiotics

Antibiotics are necessary in calves below 3 months of age to overcome certain stress conditions developed due to clinical or sub-clinical type of scour. These drugs are usually unnecessary in calves above 3 months as by this time rumen starts functioning and antibiotics are liable to interfere in normal functioning of ruminal microbes.

Feeding Milk Replacer

The objectives of feeding milk replace to calves are primarily to reduce the cost of raising buffalo calves and also to save milk for human consumption. Milk replaced may be fed as detailed in Tabular form. It may be noted that there is a gradual decrease in milk and corresponding increase in milk replaced at the rate of 200 g milk substitute per kg milk reduction.

The general principle to be followed for calculation of milk requirement is that *it should be half kg. less than one tenth of the body weight*. It is also necessary to give a minimum of 1 litre of milk per day with milk replaced and the mixture should be diluted about six times with warm water to provide a drink in suspension form.

DISEASE

The most prevalent diseases of buffaloes are much the same as those of cattle, viz. rinderpest, foot-and-mouth disease, haemorrhagic speticaemia and anthrax. Buffaloes are generally less susceptible than cattle to foot-and-mouth disease but, when contracted, its debilitating effect may be disastrous: an attack during growth may cripple the animal for life, and if it occurs soon after calving often the entire lactation is affected.

Buffaloes are much less susceptible than cattle to rinderpest, when managed under comparable conditions; the Egyptian buffalo is reported to be immune to natural infection. They are often infects with trypanosomes but the disease runs a chronic course giving rise to no visible symptoms.

They are also infested with most of the larger parasites of cattle but the extent of the damage caused is often less noticeable. As would be expected, the incidence of liver fluke is higher than in cattle. Of the arthropod parasites, the stable fly (*Stomoxys calcitrans*) and the so-called buffalo fly (*Siphona exigua*) attack buffalo more than cattle. On the other hand buffaloes are less affected by ticks but, as calves, they may be sorely tried by lice and mange mites.

9

SHEEP

Sheep are maintained in the tropics mainly for the production of meat, their other functions being to supply skins, milk, manure, and hair. There is practically no demand for wool within the region and, in fact, a striking feature of indigenous tropical sheep is that they carry either a completely hairy coat or one which has such a small proportion of wool that it has no spinning and little felting quality.

On the other hand, the fine wool produced by Merino sheep in tropical Australia is of considerable economic importance and the amount of it sold annually indicates that, given similar economic and marketing facilities, it could be produced in worthwhile quantities elsewhere in the tropics. An estimate of the world's sheep population in tropical and non-tropical areas is shown in Table elsewhere in this chapter.

Table 9.1: Estimated sheep population in tropical and sub-tropical areas.

Tropical and sub-tropical	*Thousands*	*Non-tropical*	*Thousands*
		Europe	135,200
North & Central		U.S.S.R.	133,014
America	6,848	North America	34,231
South America	43,192	South America	77,253
Asia	71,653	Asia	84,270
Africa	73,893	Africa	61,427
Oceania	50,913	Oceania	150,241
Total	176,501	Total	665,636

Total World population—832 million.

The order of importance of sheep products varies according to the countries concerned. In many districts of southern India, for instance, sheep are kept for their manure which is considered to be more important than meat or skins; the hair is unused. Some African tribes consider milk to be the chief product after meat; and with others, fat is the primary product.

In Tchad it has been demonstrated that by the use of Karakul sheep, 'fur' can be produced for export. The different functions that sheep perform have been made possible by the modifications assumed by different breeds in the process of adaptation to a variety of climatic and geographical conditions.

It is not known where or when domestic sheep originated. They have many wild relatives, but it is probable that the Argali (*Ovis ammon*) of Central Asia, the Urial (*O. vignei*) also of Asia and the Mouflon (*O. musimon*) of Asia Minor and Europe all contributed to modern breeds. Africa has no wild species except the aberrant Arui (*Ammotragus lervia*).

Sheep were probably domesticated before the ox and the domestic species (*Ovis aries*) spread in a fashion very similar to that of cattle, which has been briefly described in other chatper of this book.

The European species, having to contend with rain and cold, adapted itself by a change in coat whereby the hair fibre assumed the characters of wool, namely, non-medullated and having an imbricated formation of the superficial cells or 'scales', a crinkling or waviness constituting what is called 'crimp', and an ability to 'hold' extracellular moisture. The 'hairy wool' of tropical sheep may be over 70 per cent medulated : the best carpet wool is between 20 and 40 per cent medulated.

Species inhabiting countries where they were liable to a seasonal scarcity of food developed a reserve of fat. Where severe cold might be encountered, the fat was laid down around a short tail or on the rump and the change of a hair to a wool covering also occurred, the completeness of the transformation depending upon the degree of coldness which had to be endured.

Sheep exposed to seasonal food scarcity in warm climates, while developing the fat reserve on or about the tail, adapted themselves so as to increase their ability to lose body heat by retaining, partially or completely, the hairy coat and the comparatively long legs of their ancestors and by increasing the body surface by enlargement of such appendages as the tail, ears and dewlap.

Wool is an effective insulator and prevents body heat loss, but it also affords protection against intensive solar radiation and thus against heat gain. Hair, while protecting against solar radiation, does not hinder body heat loss to the same extent.

Those sheep which inhabited hot countries where the food supply was constant, changed in a similar way but developed no fat reserve.

All gradations of these changes are exhibited in the differences between one domestic breed and another, but the differences are not always found to take place from region to region in the way which climatic conditions alone would demand, because man, by transporting breeds, has introduced varieties of sheep into areas where they are not always acclimatized but in which, nevertheless, they not only survive but, to a certain extent, thrive.

The distribution of the various types of sheep throughout Africa illustrates the effect of man's interference. It appears that the earliest domestic sheep to be introduced were the long-legged, long-tailed, hairy type which were forced out of Egypt by the later entry of a fat-tailed, coarse-wooled variety.

While the latter spread to the south over territory in which sheep with a reserve of body fat were at an advantage, they never penetrated beyond such country to any great extent in the upper Nile valley and not at all to the Congo where the thin-tailed sheep remained.

The thin-tailed, hairy sheep likewise spread into and were found to be irreplaceable in the higher-rainfall areas of the west coast of the continent. Along the north-eastern coast, in Eritrea and Somaliland as well as in Ethiopia, Kenya and Tanganyika the fat-rumped, hairy sheep introduced from southern Arabia are now found with the fat-tailed variety.

The long thin-tailed type carried south with the Bantu also met the fat-tailed sheep of the Hottentot and those of direct Asiatic origin from which the Ronderib Afrikander were evolved. The introduction to the Cape of the Blackhead Persian in the seventeenth century proved most successful, for this fatrumped sheep, of the Somali or Hejazi type, in its improved form, extended rapidly throughout the Cape and its crossbreds returned to the tropics through the Rhodesias.

In the early part of the seventeenth century, the long-legged, hairy sheep of the West African coast were carried to Brazil, the Guineas and West Indies where they established themselves. The Iberian breeds which thrived in sub-tropical regions of America never became acclimatized to the tropical areas, in contrast to the successful introduction of the Merino into the dry tropics of Australia.

The unassisted graduated change from the fat-tailed and wool breeds of the temperate zone to the thin-tailed, hairy breeds of the tropics can be observed in Indian sheep as one passes from the northern latitudes to the southern end of the peninsula.

There is ample evidence that the power of adaptation of sheep of different breeds varies considerably, but sheep, perhaps more than any other species of domestic animal, are sensitive to change in environment. They thrive only within homoclimes and even within that limit it would appear that they have more difficulty in adjusting themselves to changes in diet and exposure to new diseases than have other kinds of livestock.

Sheep are basically fitted for life on open uplands free from forest and on short-grass prairie or steppe.

TYPICAL BREEDS

No tropical breed has been developed to the same degree as have, for example, the British breeds. There are, however, a few of the innumerable breeds and types which have been evolved in the tropics which are specially successful in their present environment and are potentially valuable because of the beneficial effect they might have on other breeds in other areas; of these, the following are noteworthy.

In Asia the *Lohi* typifies an excellent mutton and milk type. This polled, thin-tailed breed has its home in the riverine areas of the western Punjab where it is reared on fallow arable land or on the Thar desert tract. The average body weight of the fullgrown rams is about 150 pounds (68 kg.) and the average height some 31 inches (78.7 cm.) : the average weight of the ewes is 81 pounds (36.7 kg.).

It has a rather heavy, reddish-brown head with very large lop ears, a rather short, thick neck joined to a deep, wide body which is covered with a white, coarse-wool fleece. The shoulders are quite fleshy, the hind-quarters broad, the legs thick and muscular but with rather long shanks. The ewes have a well developed udder and are said to yield up to eight pounds (3.6 kg.) of milk daily at the peak of production. They are prolific breeders, respond to stall feeding and, by Asiatic standards, fatten quickly.

The sheep of the hairy-coated *Nellore* mutton breed conform to the requirements of the arid tropics in that they are little affected by high ambient temperatures and can make the most of what grazing is available. They are long-legged and measure some 30 inches (76.2 cm.) high at the withers. The rams are horned and quite heavily maned. The body is deeper and broader than the general leggy appearance would suggest: the average girth measures 33 inches (83.8 cm.). The castrated males fatten easily and produce mutton, reputed locally to be of good quality. The ewes are late in maturing and not very prolific.

In Arabia, the *Hejazi* is an important breed because it is a type common throughout Arabia and the Horn of Africa and because these sheep withstand the severest arid tropical conditions. They are fat-tailed and carry a fleece which is predominantly hairy.

The rams have a clearly defined throat ruff, measure no more than 26 inches (66 cm.) high and weigh only some 70 pounds (31.8 kg.). The body, although small, is rather stocky but the chest is narrow and the legs rather long and thin. They fatten easily but the mutton is of poor quality.

In South Africa, as has been indicated, the *Blackhead Persian* has been intensively bred and developed as a meat producer. It has been used more than any other breed for the improvement of other African breeds so that the type is found in East Africa and the arid territories of Central Africa.

It is the largest of the fat-rumped sheep and has a hairy coat, white in colour except for the black head and neck. It is polled, has semi-lopped ears and a very well developed dewlap. The neck and trunk are heavy; the limbs long but well muscled; the croup is noticeably higher than the withers.

The rump is covered with a large, thick cushion of fat which extends over the base of the tail. One of the chief characteristics of the breed is the propensity to lay on badly distributed body fat, but this can be countered by crossbreeding with the *Dorset Horn* to produce the *Dorper*. The sheep can be reared in arid conditions where few if any of the improved breeds would survive.

In East Africa, the fat-rumped *Somali* sheep closely resemble the Blackhead Persian : the tribes value them highly, breed them with care and keep them pure. Their live-weight is about 93 pounds (42.3 kg.) and their carcase weight may be as high as 75 per cent of that.

The *Masai* sheep of Kenya and Tanganyika represent another type. They are fat-tailed, coarse-wooled and dark brown. The body is quite large but rather flat sided and the legs are long.

The sheep of the Sudan are of two types, the large, long-legged *Desert* breed with a hairy coat which is either whole brown or pied. They have very long, lop ears and a long, tapering tail which, when the animal is in good condition carries much fat evenly dispersed down both sides; the adult rams then weight up to 150 pounds (68 kg.). The ewes in full milk yield five to six pounds (2.3 to 2.7 kg.) daily. The other type of Sudan sheep is the *Nilotic.* They are short-legged, have a fine, short fleece mostly white with black and tan patches; and may weigh only 25 pounds (11.3 kg.) when fully grown. The ewes give little milk.

The *Gezira* sheep which in conformation are of the Desert type resemble those found in riverine areas right across the continent from western Abyssinia to northern Nigeria where they are represented by the *Fulani* or *Uda* breed, a large sheep some 33 inches (83.8 cm.) high. The *Zaghawa* belonging to the nomadic Arabs of the Sudan are smaller and have a long, black hairy coat.

The *West African Dwarf* sheep is a type found throughout the region outside the forest belt. It has a coat of white, short, fairly fine hair, a well developed neck ruff, a long, thin tail and rather long legs in proportion to its height. The average adult weight is about 80 pounds (36.3 kg.) and in a well-fed flock of lambs it was found to be 68 pounds (30.8 kg.) when they were six months old.

In the Western hemisphere, the established *Criollo* have been improved by the use of such specialized meat breeds as the *Corriedale, Romney Marsh* and the English Down breeds in places where the climate allows: the *Merino* and the *Rambouillet* are used in the drier areas.

On uplands throughout the tropics, where the climate is more or less temperate, progressive owners rear pure-bred stock of the temperate breeds or use pure-bred rams with local ewes to produce crossbreds. In less favoured areas Blackhead Persian rams are used on indigenous stock.

BREEDING

One ram can serve or '*tup*' up to 60 ewes in a season but generally very many more rams are allowed for ewes that breed all the year round, where the rams are undernourished and where the ewes are dispersed. Under such conditions the average may be as many as one to six ewes as it is in India, one to ten as it is in arid regions of Africa, or one to 20 as in Arabia.

Normally the first strum occurs when the ewe is about a year old and it recurs at intervals of 14 to 19 days. In tropical sheep it occurs about 40 days after lambing. High ambient temperature appears not to affect the incidence of rostrum but it may adversely affect the incidence of gestation.

Hafez fcund that a sub-maintenance diet had no effect upon the onset of the breeding season but that oestrum was not accompanied by the usual physical signs and that the rate of conception was reduced. On the other hand there is no reason to believe that the effect of high feeding just before and during the breeding season is different from that on ewes of the improved breeds, namely, increased ovulation. Where there is a seasonal flush of grass, the best results are obtained by confining the breeding season to the months which allow the lambs to take advantage of it; and that generally is the season when most mating naturally occurs.

The period of gestation, as far as most of the indigenous tropical breeds are concerned, is not accurately known but from what records there are available it appears that it falls within the same limits as that of the temperate breeds, namely 140 to 160 days.

When there is an assured good supply of food for the ewes multiple births are favoured but, otherwise, they are not economic because the chances are against the survival of an ill-nourished lamb. Experimentally it has been demonstrated that the lambs of ewes subjected to an ambient temperature of 112° F. (44.4° C.) were significantly smaller than those not so treated.

PRODUCTIVITY

Meat

The majority of indigenous owners allow indiscriminate breeding, being content with an increase in the size of their flocks regardless of individual quality. Meat being the main commodity, milk is usually of interest only in so far as it is associated with meat.

Breeds of tropical sheep owe their existence, as do cattle, to their ability to survive periods of drought and semi-starvation. The economy of their living is the opposite of that which results in rapid growth and early maturity.

Judged by the standards of the temperate zone, they respond poorly to good feeding and; on their normal grazing, grow comparatively slowly and never become very fat. The lambs respond to good feeding more by an increase in vitality than size, and as adults or near adults most of the feed above that required for maintenance is used in the production of body fat which does little to improve the quality of the meat and, in the living animal, may be a misleading indication of it.

Crossbred progeny of ewes of improved breeds normally grow well until they are weaned but often fail to maintain their advantage later, even when the feed is locally considered adequate. The carcase quality, however, is better than that of pure-bred indigenous sheep, and although the crossbred may be smaller, it can be finished for slaughter at an earlier age.

The figures recorded by French shown in Table 19 give a good indication of what the average rate of growth in Africa may be, that is, about a quarter of the rate expected from good stock on pasture in a temperate climate. Singh (1962) recorded the following average daily gain of Indian Bihar, Bikaneri, and their crossbreds respectively: 0.178 pounds (0.081 kg.), 0.182 pounds (0.083 kg.), and 0.213 pounds (0.098 kg.).

In Rhodesia where crossbreeding with exotic breeds is practised, it is recorded that male lambs born on the Grassland Agriculture Research Station from October to November to half or threequarter bre-Blackhead Persian x Dorset Horn ewes and a German Merino, weighed on the average at weaning 73 .5 and 64 .5 pounds (33.3 and 29.3 kg.) respectively, and the females 64.1 and 61.7 pounds (29.1 and 27.8 kg.), but that growth after weaning was very slow.

The dressed carcase of sheep of pure indigenous breeds which have received no special feeding average some 44 per cent of the live-weight: Epstein records that the dressing percentage of Hejazi sheep is only 35 to 40 but, on the other hand, the dressed carcase of Somali sheep may be well over 50 per cent of the live-weight.

Table 9.2: Growth rate of sheep in Tanganyika.

	Breed		
Age	*Three-quarter grade Black-head Persian*	*Pure short-tailed Masai*	*Pure long-tailed Ugogo*
	lb. (kg.)	*lb. (kg.)*	*lb. (kg.)*
At birth	5.8 (2.63)	6-6 (2-99)	4-7 (2-13)
1 month	12.8 (5.81)	13-9 (6-30)	10.1 (4.58)
3 "	21-9 (9-93)	26.2 (11.88)	17-9(8-12)
6 "	32.3 (14.65)	38.4 (17.42)	32.3 (14.65)
9 "	39.5 (17.92)	50.7 (23.00)	36.6 (16.60)
12 "	46.3 (21.00)	59.4 (26.94)	42.1 (19.10)
18 "	55.6(25.22)	76.5 (34.70)	54.7 (24.81)
Average gain per day	0.103 (0.047)	0.142 (0.064)	0.100 (0.045)

Milk

In some countries milk production for home consumption is important: many Indian communities and most nomadic sheepowners derive an important part of their diet from sheep milk but only exceptionally is sheep milk sold on the liquid milk market. Lohi ewes, which are said to give up to one gallon (4.5 litres) of milk per day, have been cited as deriving their importance to a large extent from their milking capacity but the Lohi, like other milk breeds, have been subjected to no intensive development and the real potentials of tropical breeds are not accurately known.

That the ewes of many unimproved breeds do have the capacity to produce relatively large quantities of milk is suggested by the success attained in Israel where in a period of 20 years the milk yield of ordinary Awasi ewes was raised from an average of 273 pounds (123.8 kg.) to four times that amount by intensive husbandry and selective breeding, an amount exceeding that of many European breeds.

The Lacuna, a dairy breed of southern France, yields an average of 171 to 295 pounds (77.6 to 134 kg.). The Dwarf West African ewes, under conditions where a high yield would not be expected, average 88 to 110 pounds (40 to 50 kg.) in 120 to 135 days, and when well fed the yield is as much as 165 to 187 pounds (75 to 85 kg.). The Sudan Desert ewes give up to five to six pounds (2.3 to 2.7 kg.) daily.

Cheese of different kinds is commonly made from sheep milk for local and immediate consumption. Ghee is made in considerable quantities in Gujarat in India and in some surrounding districts.

Ewes are usually milked only once daily but they respond to management much in the same way as cows do.

FEEDING

Practically all tropical sheep are maintained on grazing unimproved pasture and under these conditions calculations of nutritive standards are not of immediate practical use. It has been shown that the standard basal metabolism of sheep, like that of other animals, is influenced by their nutritive level, and under drought conditions tropical sheep have been maintained indefinitely on a diet 85 per cent below that considered necessary in the temperate zone.

Good feeding is specially important during pregnancy, or at least during the latter half. After such treatment ewes produce strong lambs and milk well. Generous feeding of wool sheep results in heavier fleece but of coarser fibre.

A sheep in a familiar locality feeds selectively, usually on the finer varieties of grass, on legumes and on a fairly wide variety of other low-growing vegetation; when it is transferred to a locality bearing unfamiliar herbage it appears to have little instinctive knowledge of what is suitable, and the effect is often initially detrimental.

Most tropical sheep-grazing is characterized by periods of plenty alternating with times of scarcity. In the latter seasons not only is feed scarce but it is lignified and innutritious, and it is more on that account than on the inability of the sheep to consume sufficient material that loss in live-weight occurs. The average daily consumption of pasturage of sheep in South Africa was reported by the Agricultural Department to be:

Breed	*Lb./kg. dry matter intake per 100 lb./kg. live-weight*
Merino	2.52 (1.14)
Blackhead Persian	2.07 (0.93)
Dorper	2.80 (1-27)

The difference between the Blackhead Persian and the Dorper should be noted as giving an indication of the lesser food-consuming capacity of tropical breeds.

When stall feeding is practicable, if time is of secondary consideration, the feed may consist entirely of succulent fodder such as sweetpotato vine and groundnut haulm, a method followed by the Kikuyu of Kenya and others, to their satisfaction.

If more intensive methods are necessary a gradual increasing amount of up to threequarters or even one pound (0.34 or 0.45 kg.) a head per day of concentrates should be fed in addition. A mixture of equal parts of a cereal-maize is particularly suitable-and the bran of wheat, sorghum or rice with about 15 per cent by weight of an oilcake such as cottoncake is satisfactory.

When handfeeding is necessitated by a lack of grazing, legume hay is the best fodder to supply as it provides all the requirements of even pregnant ewes, whose need of adequate nourishment is the greatest. Sheep can be brought through a prolonged period of drought on hay made from fortuitous vegetation with no supplement, but if poor hay is enriched with a little protein concentrate at the rate of about quarter of a pound (0.11 kg.) a head each day, normal growth and development can be maintained.

Salt in one form or another should always be available, the minimum requirement being a quarter of an ounce (7 gm.) per day for each sheep. Other minerals need be supplied only when there is a known or suspected deficiency in the food.

Sheep on good grazing may not drink for many days at a time even in summer, but it is advisable that they should have regular and frequent access to water, and when they are on dry feed water should be supplied *ad lib.*

ESTIMATION OF AGE

It is often assumed that the teeth of sheep of early-maturing breeds erupt at an earlier age than those of others but what evidence is available indicates that this is not the case. On the other hand, the teeth of the faster-growing sheep of a-breed erupt at an earlier age than do the teeth of the slower-growing sheep of the same breed.

In the absence of accurate records of the age at which sheep of specific tropical breeds cut their teeth, the detailed observations recorded by Starke and Pretorious concerning Blackhead Persians and their crosses may be taken as a rough but reliable practical guide. Their tabulated conclusions are reproduced in table elsewhere in this chapter.

The age range of eruption includes the period from the appearance of the first tooth of each pair until the next pair erupt.

Table 9.3 : Ages of sheep according to permanent incisors.

Incisor stage	*Possible age Months*	*Most likely age Months*
Temporary	Up to 18	Up to 14
Two-tooth	12-27	14-20
Four-tooth	18-33	21-25
Six-tooth	24-45	26-32
Eight-tooth	above 28	above 32

The authors point out that the only reliable evidence that can be deduced from these investigations is that a sheep with two broad incisors is 14 to 20 months old and that a two- to three-year-old sheep can be either four- or six-tooth (usually indicating the age of 21 to 32 months); and that a three-year-old sheep can be expected to have its full complement of eight'. Beyond this age the teeth are gradually worn down.

WOOL PRODUCTION

Australia and East Africa are the only countries with major aggregations of wool-growing sheep in the tropics. In East Africa woolgrowing sheep occupy the highland areas at an elevation of 5,000 to 8,000 feet (1524 to 2439 m.). These have a fairly reliable rainfall.

The main aggregations of sheep in tropical Australia are in the States of Queensland and Western Australia. About 7,000,000 Merinos or one-third of Queensland's flock are in the tropics. Western Australia has less than one million Merino sheep in her north-western tropical areas. In each state sheep raising is confined to the semi-arid pastoral districts. Here the average rainfall varies between 10 and 20 inches (25.4 and 50.8 cm.) a year.

However, it is extremely unreliable and poorly distributed. When they occur the summer rains may be very intense-three inches (7.6 cm.) per hour have been recorded. In some years 40 or more inches (101.6 cm.) fall during the first four months; in others as few as five or six inches (12.7 or 15.2 cm.) fall in all 12 months. Winter rains occur infrequently.

The indigenous grasses in tropical Australia are tussocky and are usually sparsely spaced. In Queensland *Astrebla spp.* predominate while *Triodia spp.* predominate in Western Australia. The unreliable rainfall and sparse vegetative cover combine to give low stocking rates. In tropical Queensland they vary between four and ten acres (1.62 and 4.05 hectares) per sheep; in tropical Western Australia they may be as low as 15 or 20 acres (6.07 or 8.10 hectares) per sheep.

In these circumstances extensive methods of husbandry are practised. Average sheep properties in tropical Queensland are between 30,000 and 40,000 acres (12,000 and 16,000 hectares) in area; some are as large as 120,000 acres (48,000 hectares). In tropical Western Australia they are even larger.

In tropical Australia Merino sheep predominate. They usually grow long-stapled high-yielding wool with a spinning count of 64' s. to 70' s. In tropical Queensland the average fleece weight decreases from about 8.00 pounds (3.6 kg.) near the Tropic of Capricorn to about 6.75 pounds (3.06 kg.) per head at 20° latitude south.

At the same time staple length decreases. Similarly the sheep in the more northern areas are smaller and usually weigh less than those nearer the Tropic. Low reproduction rates are common to flocks in both tropical Queensland and Western Australia. In some parts of tropical Queensland average lamb-marking percentages (percentages of lambs marked to ewes mated) are between 35 and 40. In these districts death rates are frequently high; this is especially so during drought years.

As a result flocks in tropical Australia experience difficulty in maintaining themselves. Increased sheep numbers in tropical Queensland invariably result from retaining old sheep, *i.e.* those over seven or eight years of age rather than from high lamb-marking percentages. These old sheep are a bad 'drought risk' and heavy losses invariably occur when drought conditions prevail. In tropical Western Australia reproduction rates are so low that many flocks are unable to maintain themselves and the sheep population has decreased during the last few years.

Flocks in East Africa seem to maintain themselves better than they do in tropical Australia. There are probably two main reasons for this. Because they are elevated the sheep lands of East Africa enjoy heavier and more evenly distributed rainfall and a cooler climate.

Much of the sheep pastoral country in Australia experiences air temperatures of over 95° F. (35° C.) for six or more months each year. Workers within the University of Sydney showed that exposure to air temperatures of 95° F. (35° C.) or over was inclined to render Merino rams temporarily infertile. Similar observations had been

made in Missouri in the United States. Here it was suggested that the hot weather decreased the activity of the thyroid gland. This in turn decreased spermatogenesis. The American workers claimed that the administration of thyroid extract during hot weather prevented the occurrence of so-called 'Summer Sterility' of rams.

Workers within the Sheep and Wool Branch of the Queensland Department of Agriculture were unable to repeat the results reported from the United States. On the other hand they showed the intratesticular temperature of Merino rams to be 6-8° F. (3.3-1.4° C.) below rectal temperatures.

More recent work at the Ian Clunies Ross Laboratory, New South Wales, has shown that an exchange of heat occurs between the incoming arterial blood contained in the long but greatly coiled internal spermatic artery, and the outgoing venous blood contained in the pampiniform plexus, which is closely adherent to the artery. Thus the testes of the Merino ram are supplied with precooled arterial blood.

This mechanism depends in the first place upon the sweating capacity of the scrotum, which varies between rams. The semen of rams, which are able to maintain a large gradient between subcutaneous scrotal and rectal temperatures, appears to be less severely affected than that of rams whose subcutaneous scrotal temperature approximate rectal temperatures when the animals are exposed to high air temperatures.

Workers at the University of Sydney also showed that insufficient carotene intake caused temporary infertility of rams. Owing to the seasonal nature of the rainfall the native vegetation soon becomes dry and lifeless. This means sheep may experience long periods when their carotene intake is too low to maintain the necessary levels of vitamin A. On the other hand workers in the Sheep and Wool Branch within the

Queensland Department of Agriculture were unable to show that vitamin A supplements for pregnant ewes resulted in any increase in lamb-marking percentages under field conditions.

At the same time heavy losses amongst new-born lambs were demonstrated under field conditions in tropical Queensland. Over 36 per cent of all lambs born died between birth and marking. Some of these losses were due to exposure, some to predators and some to weakness coupled with low birth weights. Ewes pregnant during the summer when the grass was green produced lighter lambs (average birth weight of five pounds [2.3 kg.]) than those pregnant during the winter.

Studies on the effect of high atmospheric temperatures upon pregnant ewes soon showed that Merino ewes constantly exposed to high temperatures during pregnancy produced smaller, weaker lambs than those exposed to the cooler temperatures that prevail during winter. This finding has important practical implications.

In tropical Australia the differences between the hours of daylight at the summer and winter equinox hardly exceeds 2 hours. Here Merino ewes do not exhibit the usual strict seasonal patterns of cestrum and anoestrum. In addition the presence of freshly mated rams in the ewe flock will excite them to cestrum during the spring when the light environment is increasing.

These circumstances permit the mating of ewes during the early summer, to lamb in the autumn on green feed anticipated from the summer rains. However, this practice is often associated with extremely heavy lamb losses.

Parer (1963) found that the rectal temperature and respiratory rate of shorn sheep was significantly higher than that of unshorn sheep when exposed to high ambient temperature. He also noted that on exercise the body temperature of shorn sheep rose higher than did that of unshorn sheep and that sheep in poorer body condition were more thermolabile than those in good condition.

NUTRITION

The nutritional problems associated with the husbandry of sheep in tropical Australia fall into three distinct classes. Firstly there is the problem of absolute drought. This occurs when summer rains fail completely or as the result of two or three successive years when the summer rain is below average.

Heavy stock losses may occur. Lambings fail and the clip is light and dusty. Severe droughts occurred in tropical Queensland in 1902, 1914, 1919, 1925-27, 1935, 1939, 1945-48 and 1952. Drought mitigation is a difficult problem. The indigenous pasture may be harvested for hay. However, the yield per acre is low and the cost per pound of S.E. and of D.C.P. conserved is high. Further the terrain is rough and machinery repair costly.

Since 1953, crops which may be conserved as silage have been grown with some success. Grain sorghum planted in ground that has been in fallow for a full 12 months has proved most successful. Yields of up to 12 tons per acre (30 m. tons per hectare) have been obtained; the average has been a little over five tons per acre (12.5 m. tons per hectare).

The green crop is harvested mechanically with a forage harvester and conserved in pits each containing about 120 tons. Further studies are needed to determine the optimum size of storage pits.

Silage of this kind has a low protein content-usually about five to six per cent. In feeding trials it maintained sheep quite well when fed at the rate of two to three pounds (0.91 to 1.4 kg) per day plus two to three ounces (57 to 85 gm.) of meat meal per head per day.

Secondly there is the problem of partial drought and mineral deficiencies. Following the summer rains the grass grows quickly. At this stage its protein content may be between 12 and 15 per cent. Within four months seed is set and the protein content of the standing herbage has fallen to as low as three per cent.

This is insufficient to ensure maximum wool production and normal reproduction. Unfortunately the grass may remain in this condition for the ensuing six or eight months-except it becomes drier and more heavily lignified. This rather inadequate nutritional level produces some of the most outstanding wools grown in tropical Australia. They are known as the 'North Queensland Specialities'.

These are scoured wools, which are extremely soft and fine. Incipient copper deficiency is by no means uncommon amongst sheep in tropical Australia. In itself this condition is something of a paradox. There is adequate copper in the soil-frequently up to 50 p.p.m. Plant copper may attain a maximum of five p.p.m. on a dry matter basis.

Once the flush growth has passed it may be as low as two p.p.m. Sheep suffering from copper deficiency cut lighter fleeces and their wool lacks tensile strength and elasticity. In the circumstances that prevail in tropical Australia the correction of copper deficiency is difficult. As worm parasites do not occur, the sheep are not drenched; in fact they are not handled frequently.

The sheep do not eat cuperized salt licks readily and high freight charges make these expensive. Some success has been obtained from injecting soluble copper salts or alternatively drenching with solutions of copper sulphate when the sheep are periodically handled for shearing and such-like operations.

Toxicity resulting from excesses of fluorine and perhaps of selenium constitute the third major nutritional problem. Most of the sheep pastoral country in tropical Australia draws artesian water from underground sources. In some instances this contains sodium fluoride-perhaps between three and five p.p.m. This water is reticulated through long, slow-flowing open drains.

The normal evaporation from a free water surface in much of this country is about 100 to 120 inches (254 to 305 cm.) a year. Therefore, the fluoride content of the water in open drains increases quickly. Levels as high as 10 or 15 p.p.m. are soon attained. Sheep drinking this water are adversely affected and show clinical symptoms of fluorosis.

The wool on the back of sheep in tropical Australia becomes dry and brittle. This condition is known as 'warty tip.' It is caused by the intense radiation of the tropical sun. This divides the double sulphur bonds in the cystine molecules, leaving the fibres weak and perished. The other environmental conditions such as mean temperatures, the amount of feed, distances sheep have to walk for feed and water, distribution of rainfall all combine to produce distinctive types of wool in the different districts.

The main variations are in fleece weight, staple length, commercial count as judged by crimps per inch and fibre diameter. These factors also combine to influence reproduction rates and the general husbandry practised, *i.e.* whether breeding flocks are maintained or not. In this way the environmental conditions dictate many practices. The environment itself is hard to change, so the main problem of animal husbandry is to make the animal fit the environment!

BREEDING

Increased reproduction rates are essential to maintain flocks in most parts of tropical Australia. They are also necessary to provide sufficient sheep for adequate selection to ensure genetic improvement.

The structure of the Australian Merino sheep industry facilitates the transmission to the flocks of genetic improvement achieved in the studs. Improved methods of husbandry in sheep studs have led to marked increases in reproduction rates; this, in turn, has allowed greater intensity of selection.

A series of observations revealed considerable differences in the skill of stud masters and sheep classers to select highly productive sheep. Few men were very proficient and none achieved the accuracy obtained when selection was based on an assessment of wool quality coupled with the weighing of the fleece.

Stud masters in tropical Queensland are now using fleece measurement as an aid to the selection of high-producing sheep required for stud purposes. This has resulted in twofold, and in some instances, threefold increases in selection differentials. Fortunately, the majority of the fleece characters of economic importance in Merino sheep are highly heritable.

Therefore, considerable progress can be made by mass selection-provided that the selection is accurate! Selection is simplified by the fact that wool is the only commodity produced by sheep in tropical Australia. Therefore, clean scoured wool weights can be accepted as the cornerstone for selection.

In many respects, they are a single index of the sheep's ability to produce in the particular environment in which it is placed. By examining the various factors that have contributed to clean wool weight, it is possible to direct the available selection pressure as well as to keep certain characters, such as fibre diameter, within required tolerances.

The selection of sheep for their ability to produce and rear lambs promises to be of immediate advantage, and is being practised in some studs. This character does not appear to be highly heritable. However, there is an association between the occurrence of skin wrinkle, and of 'wool blindness' and low reproduction rates. Therefore, current trends in selection are towards sheep with plainer skins and more open faces.

With the aid of fleece measurement the probable rate of genetic improvement of sheep in tropical Queensland may be trebled. Even then only one ounce (28 gm.) of greasy wool would be added annually to their average cut per head.

The suitability of different strains of sheep to the various microclimates in tropical Australia has still to be investigated fully. In the final analysis this is a matter of land utilization. It entails finding the strains of sheep that produce the greatest economic return from wool and lambs reared in the different districts. This, in turn, calls for studies in the form, habits, modes of life and physiology of sheep.

GENERAL HUSBANDRY

Because the majority of apparel wool sheep in tropical areas are run under extensive conditions they are not handled very frequently. In pastoral Australia the whole of the management programme is influenced by the time chosen for shearing. The most popular times are autumn and spring. Sheep shorn in the autumn are usually crutched in the spring and vice versa.

This ensures that the wool is short, and hence dry, during those times of the year when the blowflies that oviposit in damp wool are most likely to be active. Blowfly strike can be prevented by docking the lambs tails at the correct length, and by performing the Mules' operation on young sheep.

However, it may not be possible to practice either of these operations in countries where screw worm (*Chrysomia spp.*) occurs, and there cutaneous myiasis may be combated by impregnating the wool with such larvacides as organo-phosphorus compounds.

The breeding programme is usually arranged so that the ewes lamb soon after shearing or crutching. About two and a half per cent of rams are joined with the ewes for about 42 days, i.e. three oestrous cycles. Thus lambing lasts for about six weeks, and as soon as it is completed the lambs are ear-marked and docked. They are usually weaned about four to five months later.

As foot-rot does not occur in semi-arid regions little attention is given to the sheep's feet. Infestations of 'barber's pole worm' (*H. contortus*) may occur during or following prolonged periods of wet weather.

Body lice (*D. ovis*) do not appear to reach serious proportions in semi-arid regions. Where they do occur they can be controlled easily by either plunge dipping or by spraying the sheep with suitable insecticides.

In tropical Australia shearing is done by contract workers using machines. It is not unusual for one man to shear 120 sheep per day, while highly skilled shearers will shear twice this number. Immediately after shearing the stained pieces are removed from the fleece, which is then rolled and classed according to staple length, count, crimping, colour, softness of handle, tensile strength, and suitability for such special trade purposes as topmaking, carding, etc.

10

GOATS

Approximately 181 million of the world population of 316 million goats are found in the tropic or sub-tropical regions; more than onethird of them are in Africa and an equal number in the Indian subcontinent. Their distribution is shown in table elsewhere in this chapter.

In Europe, goats are kept mainly as suppliers of milk, and their breeding and management are almost solely directed to the improvement in their production of this commodity: the other products derived from them, that is, meat, hair and skins, are subsidiary. In other parts of the temperate zone, goats have been developed for their wool.

Table 10.1: World distribution of goats.

Tropical and sub-tropical regions		*Non-tropical regions*	
	Thousands		Thousands
North and Central America	13,140	North and Central America	3,503
South America	18,594	South America	5,856
Asia excluding	9,650	Asia	91,295
India, Pakistan, and Ceylon	70,918	U.S.S.R.	7,290
		Europe	15,900
Africa	69,144	Africa	11,395
Oceania	44	Oceania	24
Total	**181,490**	**Total**	**135,263**

Total World Population-approximately 316 million.

In the tropics, although a few breeds with a marked propensity for milk production have been evolved and are primarily maintained for that purpose, the majority owe their existence to the fact that they can thrive as meat producers in conditions in which it is difficult for other species of domestic livestock to live.

Their unsurpassed ability to forage, their preference for browsing on a wide variety of vegetation, their catholic taste for any kind of edible material, and their power to withstand the extremes of a tropical climate, account for their widespread distribution throughout the zone.

In intensively cultivated areas and in cities, where grazing land is scarce, cattle and buffaloes are costly to keep but goats, if they are dispersed, can subsist on refuse or find a living by browsing or grazing on incidental vegetation and while doing so produce cheap milk and meat.

In arid countries or in broken land unsuited for cultivation where vegetation is sparse, goats rove more widely than other herbivora and gather the browsing overlooked by cattle, the grass which is too short for cattle or, in the absence of better food, the drier coarser grazing unacceptable to sheep.

Although there are large numbers of goats in humid tropical regions, it cannot be said that they flourish there; and even in areas of moderate humidity they are at their best only on light, sandy soils or on well-drained hill grazing; nevertheless they are the meat and milk producers in many regions infested by tsetse fly where cattle cannot be maintained.

The domestic goat belongs to the genus *Capra* and is in all probability descended from the species *C. cegagrus,* although the ancestors of the wool-bearing types of goat may have been *C. falconers.*

From prehistoric times, as individuals of the species were domesticated, multiplied and spread from the country of origin in the mountains of Asia, their development conformed with the requirements of their new surroundings.

New varieties of breeds evolved bearing physical characters moulded by this environment but which in a vague way denote the chief functions of the breed and by more obvious signs of coat-colour, body size or conformation distinguish one breed from another.

Thus today, while there are innumerable breeds varying in some unessential physical detail, all can be classified from the functional point of view in three groups according to the principal commodity they produce, namely, the meat, milk and wool goats. The first two groups are geographically very widely distributed, and within each there are types of individuals which differ quite markedly in body conformation, varying from those rather large animals with long legs, fine skins and rather narrow bodies suited to desert life, to the smaller, short-legged stocky type fitted for the easier life in humid climates or restricted surroundings.

The *Cutchi* breed of Bombay, the desert type goat of the Sudan and the *Sapel* of West Africa are typical of the long-legged meat varieties. The average weight of Cutchi goats when full grown is about 80 to 90 pounds (36.3 to 40.8 kg.) but the others weigh little more than half that. The average carcase weight is about 45 per cent of live-weight, but average carcase weights of above 62 per cent have been recorded.

The short-legged varieties are typified by the *Hejazi* of Arabia, the *Dwarf West African* goat and the *Moxoto* of Pernambuco, the average live-weight of which is about 60 pounds (27.2 kg.). The carcase weight is usually of a higher percentage than that of the former variety.

In the temperate zone, the breeds of the heaviest goats are the *Anglo-Nubian* and the *British Saanen.* The average live-weight of the males is 180 to 220 pounds (81.6 to 99.8 kg.) and that of the females, 140 to 170 pounds (63.5 to 77.1 kg.).

The best-known representative of the tropical milch group are the *Jumnapari* of India and the *Nubian* of the Sudan, both of which have been used in the production of the modern Anglo-Nubian breed. The large up-standing Jumnapari (Plate 32) has a characteristic convex face, rather foreshortened muzzle and long, lop ears.

The males are 36 to 40 inches (91.5 to 101.7 cm.) high at the withers and the females are 30 to 34 inches (76.2 to 86.4 cm.): the average live-weight is 175 and 120 pounds (79.4 to 54.4 kg.) respectively. The average duration of the lactation of goats kept under good conditions in India was found to be 205 days.

The average milk yield of a recorded herd was 540 pounds (244.9 kg.) in 274 days, the highest day's yield being 8.4 pounds (3.8 kg.) and the highest yield in a lactation, 1237.4 pounds (561.3 kg.); average butterfat percentage was 5.2 and the maximum was 7.4 (Kaura, 1943). The Nubian is altogether smaller, the males being no more than 30 inches (76.2 cm.) high at the shoulder; the average milk yield is about three pounds (1.4 kg.) a day.

The short-legged milk group is typified by the Indian *Surti.* The Surti in a lactation period of about the same length as that of the Jumnapari is said to produce three to five pounds (1.4 to 2.3 kg.) of milk a day but no authentic records of production appear to have been made.

These figures refer to well-fed goats usually receiving supplementary concentrates, but such yields cannot be expected from animals which subsist by diligent foraging, as most tropical goats do; from them a yield of two or three ounces (57-85 gm.) each day throughout a short lactation period is all that is likely to be forthcoming.

The average daily milk yield of Indian goats, including the yield of good milking-goats with that of all others, has been estimated to be seven ounces (0.2 kg.) by Wright (1940) and one pound (0.45 kg.) by Sen (1953), in a lactation period of 150 days.

The British Goat Society (1957) recorded a list of 98 goats in Britain which in a year's lactation terminating in 1955 had yielded between 2,501 and 5,617 pounds (1134 and 2548 kg.) of milk; 55 of them had yielded over 3,000 pounds (1361 kg.). The average butterfat percentage ranged between 2.80 and 5.64.

The Saanen, *Toggenburg* and Anglo-Nubian have been used with success in the improvement of the milk-yielding quality of indigenous goats in many tropical countries where adequate feed throughout the year is assured, but in other countries, with a humid climate, the greatest difficulty has been experienced in acclimatizing goats of these exotic breeds.

The wool-bearing type of goats is represented by the *Angora* and the *Kashmeri;* they are not raised in the tropics. The former has been used extensively and has been much improved in South Africa and in North America so that whereas some fifty years ago the average production in a herd was about two pounds (0.91 kg.) per goat, the average now is about four times greater.

Considerable work has been done in developing both milk and mohair goats but no serious attempt has been made to improve the meat-producing breeds. In fact, the meat goat is one which lacks the innate ability to produce worth-while quantities of the other commodities; it is the unspecialized type of the species evolved through natural selection aided infrequently and casually by man.

If any goats may be considered specialized meat producers, the *Thenges* of the Kikuyu of Kenya would appear to be such, but these stall-fed castrates, valued for their prodigious yield of carcase fat, which are a feature of the Kikuyu husbandry, are derived from no particular breed.

A goat carcase never has as good a finish as that of a sheep, chiefly because its fat is not dispersed but tends to be concentrated around the viscera. Even when in heavy condition the carcase has the appearance of being inadequately fleshed. The grain of the meat is more compact and the colour slightly darker than mutton. Some tropical peoples prefer goat flesh to mutton or, in fact, to any other meat.

Although in many tropical regions goats constitute the basic meat supply of the people, government livestock departments do not encourage an increase in their numbers because the normal activities of the animals are detrimental to soil preservation, forestry and settled agriculture.

They are very destructive of trees and shrubs ; they will keep stripping the bark off trees until the trees die. Thus it is essential that goat husbandry should not be encouraged where there is any real danger of soil erosion unless the goat can be kept under

control. In certain circumstances they keep down scrub on grazing land, thus maintaining pasture, and thus too they prevent the invasion of tsetse flies where it is threatened. Their position in tropical agricultural economy is discussed by Maher.

Goats of some breeds have skins of higher quality than others, for fine leather-making, but no breed is raised specially for that purpose. The product is a valuable one; an annual revenue of over half a million pounds is raised from the export of goat skins in Nigeria where the goat population is under six million.

Management

Goats are rather difficult to hand-feed, chiefly because although they can be fed on standard agricultural produce they do best when their fodder is composed of a variety of vegetation which is often difficult to collect and troublesome to stall-feed. Moreover, their active nature demands scope and freedom of movement if they are to be maintained in perfect health.

In the tropics it is unusual to confine them except at night or when it is necessary for their safety. Where pasturage is extensive and dispersed they are more often herded with sheep or cattle or with both, than as self-contained flocks. A few goats are commonly kept with herded sheep to act as leaders and encourage wider ranging.

As has been indicated, their feeding habits supplement those of both cattle and sheep and do not necessarily compete with them. Goats do not by themselves close-crop pasture; they are both selective and wasteful. The acreage required per head depends, of course, upon the quantity and nature of the vegetation.

Goats are not easily held within fencing and nothing less than stout mesh wire, of which galvanized chain-link is the best, is really reliable, or a fence which consists of woven wire topped by two strands of barbed wire.

The simplest structural accommodation suffices for range goats. In the daytime, shade should be available when the sun is intense. At night, dry quarters are essential, with or without overhead covering, shelter from rain being one of the most important factors in goat husbandry outside the arid areas. Protection from cold may also be needed but usually a wind-break is all that is required.

Goats are particularly susceptible to the effects of bad ventilation associated with crowding in insanitary buildings: when they are so herded, high mortality from pneumonia often follows.

If the highest productivity is to be obtained from the specialized milch goat, she must be treated in the same way as the high-producing cow, that is, she must be fed according to her milk yield and she must be saved unnecessary expenditure of energy or have food to compensate. No elaborate housing is necessary but provision should be made not only for the protection of the goat but also for the handling of the milk to keep it free from dirt and taint.

The building therefore must be constructed so that it is dry and cool, is well ventilated and can be kept in a sanitary condition at all times and in all seasons. In equitable warm or hot climates, a thatched roof over a concrete floor is all that is needed, but in other circumstances the weather side may have to be walled or the quarters may have to be walled on all sides.

The roof should be at least seven feet six inches (2.29 metres) high in front sloping to five feet (1.53 metres) behind for the lean-to type of stable; but with a ridge roof the minimum height should be tenand-a-half feet (3.2 metres) at the ridge and seven feet (2.14 metres) at the eaves.

If it is impracticable to have the floor made of concrete, which is undoubtedly the most satisfactory material, rammed earth or clay may be substituted, but in any case the actual standing should slope from the front at a gradient of about one in forty to a wide, shallow open drain behind. Large goats require eight square feet (0.74 square metres) of standing, but six square feet (0.56 square metres) is ample for the smaller types.

The standing should be arranged on the plan of a modern dairy stable with the water receptacles fitted at least one foot (0.31 metres) above ground level and the fodder racks high enough to allow the goats to pull the feed down as they would do if they were browsing. When practicable, each standing may be fitted with a platform raised about

18 inches (0.46 metres) from the ground and just broad enough to allow a goat to stand and lie on it.

Alternatively the whole standing may be covered with a movable slatted floor raised a few inches from the ground so that drainage is kept free and air circulation is promoted.

BREEDING

The first oestrus may occur when the female goat is only a few months old, in fact, eight-week-old females have been known to conceive and ultimately bear normal progeny. Male goats may make an effective service when they are no more than four months old but even in range flocks subject to the barest restriction it is unusual for breeding to take place at that early age and it is generally considered inadvisable to mate before the animals are about nine months old.

In well-managed flocks breeding is not allowed until the females are a year to a yearand-a-half old, depending upon the time of the year they were born, and the males are used only lightly at one year old and brought into full use when they are two years of age. They may then serve up to 100 females. The-best time for service is 12 hours after the onset of heat.

The reproductive activity of goats, like that of sheep, is induced by the shortening length of day. In the northern regions of the world, where the length of day varies greatly according to the time of year, eestrus occurs mostly from July to October: in the tropical region, where there is little seasonal difference in the length of day, the cestrous cycle recurs throughout the year although it is most in evidence during certain seasons.

The duration of cestrus is usually 24 to 48 hours but in certain breeds, for instance the Indian *Beetal,* it may be as short as 18 hours on the average. Recurrence is at 18-21 day intervals unless pregnancy occurs.

The gestation period is from 145 to 153 days. Tropical goats commonly kid twice in one year and usually three times in two years. The average number of kids born at one parturition is generally recorded as 1.3 but in some breeds, for example the Surti, the average is as high as two, and as many as 20 per cent of the goats of some breeds give birth to three kids.

In a herd of Fiji goats, breeding occurred three times in two years with a mean kidding interval of 264.4 days. On the other hand the Nubian type of goats usually kid once a year and give birth to only one kid. The 'unimproved' goat very rarely has difficulty at parturition.

REARING

When kids are left with their dams they require no particular attention: they are strong, more active and more able to look after themselves than the average lamb. If they are bottle or bucket fed they require milk every three hours during the first week and every five hours for the next two weeks, thereafter milk feeding can be reduced to twice a day as the kid gradually takes to solid food.

Kids of the large milch type of goat take up to two pints (1.1 litres) a day but kids of non-specialized breeds seldom need more than one pint (0.55 litre). When goat's milk is required for purposes other than rearing, either cow's milk may be substituted or when the kids are over two weeks old they may be reared on a gruel made by mixing equal parts of a milled cereal and a milled legume with a solution of one part of milk to two parts of water, the gruel being warmed to just below blood-heat when being fed.

They may be weaned at six weeks old when they should be eating enough solid food upon which to exist, but it is better to continue the milk feeds until the kids are three to four months old, and those selected for breeding should not be weaned until they are five or six months old. Kids that are to be slaughtered for prime meat should not be weaned.

The first solid food if it is to be stall-fed should take the form of young, tender roughage such as sweet-potato shoots, young cassava root, bamboo shoots or legumes either green or as hay.

The male kids not required for breeding should be castrated early, that is, any time after they are two weeks old although the operation is generally deferred until the animal is between two and three months of age. Bloodless castration by the use of an elastic band, the 'elastrator', is particularly suitable for goats.

FEEDING

Goats will accept a great variety of feed, appreciate it and thrive on it, but what is acceptable to one goat is not always acceptable to others all generally refuse anything which has been soiled by other goats or animals of any species.

The basis of all goat feed should be mixed roughage adequately supplemented to meet the nutritional needs of high production. These French estimates to be approximately as shown in table elsewhere in this chapter.

The amount of starch equivalent and protein in the roughage consumed by the goat, assuming it has fed to capacity, is ascertained by consulting tables of the composition of local foods, and any deficit between that and the requirements calculated by use of the figures given in Table elsewhere in this chapter is made good by feeding appropriate amounts of a suitable concentrate or concentrate mixture to which has been added the minerals recommended for cows in milk.

Table 10.2: The nutritional requirements of goats.

Breed	*Average live-weight lb. (kg.)*	*Requirements for maintenance lb. (kg.) per day*		
		S.E.	*P.E.*	*Dry matter*
Temperate pure-breds	130 (59.0)	1.5 (0.68)	0.07 (0.032)	4.1 (1.86)
Half-bred	90 (40.8)	1.2 (0.54)	0.05 (0.023)	3-2(1-45)
Tropical pure-breds	60 (27.2)	0.9 (0.41)	0.04 (0.018)	2.4(1.09)

Percentage of Fat in milk	*Requirement for each lb. (0.45 kg.) of milk produced*	
	S.E.	*P.E.*
4.2	0.275 (0.125)	0.067 (0.030)
4.7	0.300 (0.136)	0.074 (0.034)
5.2	0.325 (0.147)	0.081 (0.037)
5.7	0.350 (0.159)	0.088 (0.040)

Sometimes a cereal such as maize will suffice; at other times a protein concentrate is required and a kibbled oil cake will answer the purpose, but a mixture is generally needed. The following one is recommended by the U.S.A. Bureau of Animal Industry:

100 parts	corn
100 parts	oats
50 parts	wheat bran
25 parts	linseed oil meal.

It should be fed as a supplement to as much good roughage as the goat can eat, at the rate of one pound (0.45 kg.) for each quart (1.1 litre) of milk produced.

At the Goat Breeding Farm at Etah, India, the following was found suitable:

4 parts	crushed barley
8 parts	crushed maize
9 parts	linseed or mustard cake
8 parts	hand-milled wheat bran.

A pound and a half (0.68 kg.) of the mixture was fed daily to goats whose average weight was 150 pounds (68 kg.) and whose average milk yield was two pounds (0.91 kg.) daily, the average butterfat content being 5.2 per cent. The roughage consumed was varied in nature but only fair in quality.

In East Africa, a mixture recommended as a supplement by Lowe (1942) is:

33 parts	crushed maize
33 parts	groundnut cake
50 parts	cotton seed.

Good roughage such as sweet-potato vines, groundnut tops or cowpea haulms alone suffice for the adult goat when not in full milk, and it is also sufficient for the growing and fattening goat. Normally goats browse rather than graze, and any range which does not provide browsing cannot be considered adequate.

Full-grown goats are fattened for good-quality meat on the best fodder supplemented with up to one pound (0.45 kg.) of a cereal such as maize, sorghum or rice. The best goat meat is derived from suckling kids slaughtered when three to four weeks old, or from young goats about five months old fed on milk and as much of such feed as bran, tubers or concentrate mixture as they can be induced to eat.

11

CAMELS

Domestic camels function mainly as a means of transport, a source of traction power and as producers of meat and milk. They are used in arid regions because of their ability to flourish on the fibrous vegetation found there, which cannot be utilized to the same extent by any other species of domestic animal, their power of endurance in prolonged drought, and their capacity to do without drinking water for comparatively long periods.

There are two varieties of camels, namely, *Camelus dromedarius,* the single-humped camel found in the tropics, and *C. bactrianus,* living in regions of intense cold. Although the two varieties are so differently adapted, the individuals of either species are not amenable to sudden change of climate, surroundings or methods of management.

There are some nine million camels in the world: two-thirds of them are located in the tropics, half of these in the east of tropical Africa. It seems inevitable that their number must decrease as mechanical transport becomes more universal.

In the marginal desert areas where the camel and the horse were the sole means of transport, the machine has virtually displaced them although the camel retains its somewhat limited use in the plough. Even in the desert the machine replaces the camel to an extent which is now apparent in certain regions, but camels maintain, if they do not actually increase, their usefulness as meat producers.

In the Sudan, Somaliland and to a lesser extent in some of the neighbouring countries, a considerable number of camels are maintained and fed specially for slaughter. Elsewhere camels butchered for meat are the worn-out, the incurably injured, the barren and the misshapen.

There are no breeds specially developed for meat production.

Conformation: Specialized breeding of the single-humped camel has produced two very different types, namely, the riding camel and the baggager, but the breeds within these types are not marked by so pronounced functional or structural differences as distinguish, for instance, those of cattle; nevertheless the camels of each country or locality carry hereditary traits which stamp one from the other in no uncertain manner, and, similarly, family '*lines*' within each breed are recognized and critically appraised by traditional camel-owning communities.

But environmental influences give rise to types which are more easily distinguished by the inexperienced than the various breeds within the type. There is, for instance, far more in common between the *Egyptian delta* camel and the riverine *Sindhi* of Pakistan than there is between the Sindhi and his immediate neighbour, the *Kuchi,* from the hills.

The small, compact, muscular, heavy-boned hill type, bred in rough upland regions is very different from the more rangy, long-legged, more loosely coupled, plains camel. The former is only some six to six-and-a-half feet (1.8 to 2.0 metres) high, while the latter may be up to seven feet (2.14 metres) high at the withers.

The conformation and performance of the plains camels vary in all degrees between the light, fine-boned, thin-skinned, alert, desertriding type, and the massive, but rather mean-looking phlegmatic baggage type from the riverine areas, accustomed to good living and regulated activities. The average Indian camel weighs 1,170 to 1,230 pounds (531.7 to 557.9 kg.), and others from 1,000 to 1,230 pounds (453.6 to 557.9 kg.) (Leese, 1927).

There are many niceties of conformation and temperament which enhance or depreciate the utility of a work camel but, provided the animal is healthy and free from injury, the only point of conformation of real importance is that the limbs should be set on straight; the feet not point outwards, the elbows cramped inwards, nor the hocks show any tendency to touch either when at rest or on the move. Such conformation should ensure a free, easy, straight gait which is the first essential in a work-animal.

A camel selected for immediate work should be in good condition as indicated by a plump, rounded hump, well covered ribs and muscled loins. It should be able to sit and rise with ease when under load. The subcutaneous oedematous condition which normally arises after a very thirsty camel has been given a large dose of salt and then has been allowed to drink its fill may temporarily make a thin camel look well covered.

Although the exotic demand for first-class riding camels has practically ceased, the best of the individuals suited for that work are still selected with care and used with great discrimination, but the majority from all categories are employed for pack transport or as family beasts of burden. Only a relatively small number are used for ploughing or for pulling vehicles on modern roads.

MANAGEMENT

The camel is an amenable, patient animal, easily trained to work. Generally females are used almost as much as males and the calf, running with its dam, is accustomed to handling from the earliest age.

At two years old he is introduced to the discipline of control by head rope or by a nose peg made of wood, bone or, very occasionally, of metal passed through the nose below and towards the extremity of the nasal bones, at first merely carrying the attachments and then having them increasingly handled and brought into use.

He is then loaded with an empty saddle and later, often by the time he is three years old, ever-increasing loads are mounted according to what he is considered strong enough to carry; judgement in this respect far too often tending to credit the young animal with more strength than he has, to the detriment of his further growth and development.

As soon as head control has been established he is induced to sit and rise at command. This he is taught by having his head pulled towards the ground while being tapped behind the knee with a stick, the word of command being continually repeated, until he kneels in the natural position preliminary to subsiding on to his hocks and finally to his breast-pad.

The best camelmen allow their females to graze at liberty till the calves are weaned. The male young-stock continue at liberty until they are about four years old. They are then handled and lightly loaded during the next year or two but are not brought into full work until they are six years, and may continue to work till they are twenty years old.

Castrated males make the best workers, provided they have not been castrated before they have attained almost the full size of the entire male, that is, at four to six years old.

Castrates are seldom able to carry as heavy loads as the entires but their work is never interrupted by the *rut* period nor are they distracted by females.

CAPABILITIES

The riding camel can carry two men but generally he is ridden by one man who carries with him up to 120 pounds (54.4 kg.) of baggage. With this burden a pace of six miles (9.7 kilometres) an hour can be maintained on level ground, and a distance of 30 miles (48.3 kilometres) covered each day for prolonged periods.

With a single rider and otherwise unburdened, a good riding camel can go at the rate of nine to twelve miles (14.5 to 19 .3 kilometres) for an hour or so and can travel fifty miles (80.5 kilometres) a day for a fortnight.

The baggage camel travels at a walking pace at just over two-anda-half miles (4.0 kilometres) an hour and can carry a full load for 15 miles (24.1 kilometres) a day for an indefinite period. What comprises a full load depends upon the strength of the individual camel and may be from 350 to 650 pounds (158.7 to 294.8 kg.): as much as 1,200 pounds (544.3 kg.) may be carried for short distances.

When camels were used in the British army, the maximum load normally permitted was 450 pounds (204 kg.) for 20 miles (32.2 kilometres) a day over prolonged periods, with one day's rest each week.

For each form of work, even for draught work, a saddle of some sort is required. Fundamentally, all satisfactory saddles are made on the same principle and consist of a rigid frame to lie along the muscular mass on each side of the animal's spine, connected by arches passing over and clear of the spine and strong enough to support the burden to be carried and to protect the spine from it: the hump lies clear between the arches.

A pad of some sort generally cushions the skin from the pressure of the frame and a restricted superstructure of padded leather makes the seat of the riding saddle, or an expanded rack-like structure gives space for baggage, or the frame itself supports the shafts of a cart. The saddle is held in position generally by a single girth, and in hilly country, by a breast rope and crupper or a neck and a tail rope.

Even the Somali 'herio' is roughly built to this pattern although it consists only of a series of mats looped together and to the camel by one long rope.

BREEDING

Females are sexually mature at three years of age but are not generally bred until they are four years old. They may continue to breed until they are over 20 years old. The average length of gestation is 370 days. Normally, only one calf is produced every two years but occasionally the female may breed twice in two-and-a-half years.

Estrum lasts for three to four days. Estrum may recur in a well-fed camel as early as one month after parturition but generally it is delayed for about a year although occasionally it recurs much sooner.

The male camel is peculiar in that, annually, he has a period of sexual activity which may last for a month or two; for the rest of the year all breeding instinct is suppressed.

His period of activity is associated with the lengthening day, that is, between January and March in the northern hemisphere and June and September in the southern.

At the equator there is no definite 'rutting' season and breeding may take place at any time throughout the year. The average duration of *rut* is about three months but it depends largely upon the animal's plane of nutrition: in well-fed camels it may last for as long as five months and, exceptionally, in old camels it may continue throughout a year.

The male may breed when he is three years old but his full reproductive power is not developed until he is some six years old; he can then serve up to 50 females and, if he is very well fed and cared for, he can cover up to 70 in a season.

While he is in *rut* his usefulness as a baggage or riding animal is greatly diminished; he loses his appetite, often suffers from diarrhoea, lacks bodily vigour, is preoccupied with an exhibition complex, is irritable, intolerant of a rival and rebellious of discipline. If he is driven at hard work, his sexual inclinations decline or disappear.

Females are often worked right up to the time of giving birth and continued in work soon after. This is not a practice which should be encouraged nor is it one which is followed by the real camel-breeding tribes for it is deleterious to both dam and calf, and is inhumane.

Females should have liberty on good grazing during at least the last months of pregnancy and for three weeks thereafter. All camels under good management are retired from work for a definite time each year, that is, usually during the rains when the conditions for work are hardest and there are the best opportunities for browsing and grazing.

REARING

The young camel is a rather delicate creature and a considerable number of them die before they are three weeks old. The colostrum, or first milk, which is known to be essential for the good health and well-being of the newly born, is considered by the majority of camel breeders to be rather a dangerous substance which should be consumed in minimum amounts.

It may be for that reason that so many new-born camels are so slow to acquire strength: it is usually at least a week before they can follow their dam while she grazes. On the other hand experience has taught that if the young camel habitually drinks to repletion, indigestion, diarrhoea and death often result.

It is customary, if the dam is a heavy milker, to hand-milk three-quarters and allow the young camel to suckle only onequarter for the first three weeks and then gradually to increase the amount allowed. A good dam should yield 20 pounds (9.1 litres) of milk daily at the peak of her lactation, and up to 6,000 pounds (2,727.6 litres) or more in a lactation.

The young are weaned naturally: the dam goes dry after about nine months, if desert fed, and up to 18 months when on richer fare. Her calf begins to graze when only a few weeks old, and the change from liquid to solid food occurs gradually and without check to growth.

DENTITION

As so much concerning the management of a camel depends upon its age, the ability to estimate the age with a fair degree of accuracy is important. The estimate is made in the usual way from an examination of the incisor and canine teeth of the lower jaw.

At birth the central pair of teeth are already erupted; at one month, the laterals appear; and by two months the third pair is through the gum. They are crowded and overlap. By one year they have grown up and are in wear.

At two years they are worn and no longer touch the neighbouring teeth. At three, they are well worn, and at four years they are quite worn down, have a square or irregular table, and are loose. By five years, the central permanent teeth have erupted, followed a year later by the next pair and all three pairs are visible at seven years old.

In the upper jaw there are two canine-like temporary teeth on either side; and three such permanent teeth are visible at about seven-and-a-half years old. There are two permanent canine-like teeth on both sides of the lower jaw to be seen at the same age.

These canines reach their full size at seven years, when they are long and still sharp pointed. As the camel ages, all the teeth become worn and blunt and by twelve years the canines are stumps and the incisors upright stubs with square or circular tables.

FEEDING

Camels prefer to browse rather than graze, and need time not only to consume their food but also to ruminate. Their mobility and lesser dependence on drinking water allow them to range over a far greater area than any other domestic animal. As a general rule, when food is fairly easily obtainable, six hours is the minimum time which should be allowed for foraging: at least another six hours are needed for rumination. They will not forage during the heat of the day if it is at all intense.

In many territories camels live entirely by browsing, sometimes mostly on low bush plants such as camel-thorn *(Alhagi maurorum)* and salt worts such as *Haloxylon recurvum.* Such camels do not normally relish grass even when it is young and fresh. In other countries, for instance in Somaliland, grass forms a principal part of the diet.

All camels under the force of necessity can be maintained in good condition entirely on conserved fodder and grain such as is given to stall-fed cattle. This method of feeding is expensive in every way and is followed only in exceptional circumstances; but a certain amount of hand-feeding is often necessary to supplement grazing and browsing in seasons of scarcity or when sufficient time cannot be spared for it.

Under any circumstances it is desirable that camels should have liberty to forage for at least six hours each day because the variety of material they consume, although far from unlimited, is generally sufficient to ensure that all their nutritional requirements are catered for. If salt is not hand-fed it is said that at least one-third of the grazing time should be on salt bush such as *Atriplex, Salsola,* and *Suaeda spp.*

Leese quotes the following as being suitable supplementary rations when 'grazing' is fair, poor or not available:

Working Indian or Egyptian baggagers

Feed	*Grazing fair* lb. (kg.)	*Grazing poor* lb. (kg.)	*No grazing* lb. (kg.)
Gram (*Phaseolus spp.*)		4(1-8)	6(2-7)
Missa bhussa (*legume straw*)		20(9-1)	30(13-6)
Millet	8(3-6)		
Tibben (*chopped wheat straw*)	8 (3.6)-12 (5.4)		
Salt		1.5 oz. (42 gm.)	1.5 oz. (42 gm.)

The standard ration for Somali or Aden camels is:

Feed	*Amount* *lb. (kg.)*
Gram or Jowar *(Andropogon sorghum vulgare)*	4(1.8)
Hay	20(9.1)
Salt	1.5 oz. (42 gm.)

The normal diet of the Somali Camel Corps camels consisted of the following in addition to grazing according to Peck.

	lb. (kg.)
Split gram (*Cicer arietimum*)	4(1.8)
Wheat bran	2(0.91)
Hay	10(4.5)
Salt	1 oz. (28 gm.)

Peck considers that the normal salt ration for such camels when working under conditions which do not cause them to sweat unduly should be five ounces (142 gm.) each day, and he found that the improvement in condition following the feeding of that amount was spectacular.

There is much evidence to show that the amount of salt usually fed to camels not consuming salt-rich plants is too little. A daily ration of five ounces (142 g.) may be considered the minimum needed to keep a full-grown working or milking camel in prime condition.

WATERING

Schmidt-Neilsen (1964) describes the camel's exceptional tolerance to heat and water deprivation and states that this does not depend on the storage of water in the body.

The body temperature of the camel is variable and when it is deprived of water, daily fluctuations may exceed 6°C. (12°F.). This physiological characteristic assists the camel in conservation of water and disposal of heat.

As body temperature rises during the day, water that would be used for evaporative cooling to keep body temperature down is unexpended, excess heat is stored in the body and dissipated during the cooler night. At the same time the elevated body temperature during the daytime reduces the heat flow from the hot environment to the animal's body and further reduces water expenditure.

The camel's very fine hair is an important barrier against heat gain from the environment; less water is utilized by unshorn camels in any given environment. Also, although the camel sweats, water evaporation is from the skin and not from the surface of the hair, and this also assists water conservation.

Camels can concentrate their urine to a marked degree, and Schmidt-Neilsen believes that they can re-circulate and re-utilize waste nitrogen under certain circumstances.

The camel can consequently stand a very considerable degree of dehydration, and in hot environment it can tolerate the loss of at least 27 per cent of its body weight, and has an enormous drinking capacity, drinking at one time, after deprivation, as much as 30 per cent of its body weight.

Nevertheless, when practicable, camels should be watered every day, when they will drink up to eight gallons (36.4 litres) : if they have been deprived for any considerable time, up to 20 gallons (90.9 litres) may be drunk by a single camel at one time.

Riverine working camels must be watered daily if they are to maintain condition; Indian desert camels need to have water every second day; others can be kept in work when they are watered only every fourth day. Camels of all breeds can subsist for remarkably long periods without drinking water, and the Somali has exceptional powers of endurance in this respect.

12

HORSES AND MULES

The proper evaluation and selection of either purebred or grade draft animals is a much more complicated job than that of properly evaluating other farm animals. In addition to the conformation, quality, and condition of the horse or mule, the buyer or judge must also consider the feet, legs, action, wind, and temperament. The importance of these last items is reflected in the old sayings "No foot—no horse" and "The feet, legs, and action are half the horse."

To properly evaluate draft animals one must first learn the proper names for the different parts, and then he must consider the purpose of work stock and how they accomplish their purpose. Draft horses and mules are designed principally to pull more or less heavy loads at a medium to slow gait. Practically all of the power for this pull is generated in the rear quarters of the horse and mule.

The pull is exerted almost entirely through the rear feet with the front feet serving mostly to support the front part of the animal as would a set of wheels. From the rear feet the force is transmitted through the rear legs, loin, back, and shoulders into the collar. These facts warrant the drawing of several conclusions:

1. The feet of a draft animal must not only be sound but should also be large so that they can obtain maximum grip on the ground.
2. The bone of the draft animal must be strong and durable and the legs must be placed or "set" properly.
3. The hocks must be broad, deep, and clean because these are the levers through which the power, generated in the muscles of the rear quarters, is transmitted to the feet when pulling.
4. The gaskins, quarters, thighs and croup of the animal must be heavily muscled, since these generate the pulling power.

5. The loin and back must be short, broad, and heavily muscled in order to carry the thrust from the rear quarters to the shoulder.
6. The shoulders must be sloping and clean cut so as to give a good seat to the collar and permit an alert carriage of head and neck and a long, easy stride.
7. The middle should be sufficiently deep and roomy to provide enough capacity for feed, especially pasture and roughage, so that the work animal can maintain his condition with a minimum of expense for concentrates.
8. The chest must be deep and wide so as to provide sufficient lung capacity to take care of the needs of the animal while working at or near top capacity.
9. The pasterns, especially the front ones, must slope at approximately a 45-degree angle, since the front feet are raised higher and, consequently, strike the ground harder than the rear feet. The angle of the pastern absorbs this shock and, thereby, tends to prevent ring-bones, splints, side-bones and other blemishes and unsoundnesses caused by excessive and repeated shock.
10. The horse or mule must be willing and alert since a balky or sluggish draft animal will waste both his time and that of the driver.

One of the most difficult things for beginners to learn is the proper placement of feet and legs. Animals are seldom seen with all four legs squarely in place. Show animals having faulty set to legs or hocks are usually placed so as to make these defects less obvious.

Once the proper placement of feet and legs is learned, however, it is relatively easy to spot defects. As an aid to learning the correct placement for the feet and legs of a horse or mule study Figure, which shows both the correct placement and the most common deviations from this.

Ideal Draft Animal

Even though horses and mules do not closely resemble each other, they serve the same purpose; their ideals in size, conformation, quality, and action are very similar. The main difference lies in the closeness to which each approaches this ideal. Some mule men contend that a mule that looks too much like a horse lacks the durability expected of a mule. There is some truth in this argument.

However, in both the sale ring and the show ring the mule is measured against much the same standard of perfection as the horse.

He is usually most deficient in quality of head; muscling over the back, loin, and croup; fullness in the rear quarters; size and quality of feet and bone; and set of hind legs.

Size

The ideal size for a draft animal will vary with the section of the country, the size of the farm, and the job that is to be done. The extra-large, clumsy horse that was popular in the North some 20 years ago, has now been replaced with a 16-hand (a "hand" is four inches), 1600-pound horse that is no taller from the sole of his foot to his chest floor than from his chest floor to the top of his withers. This newer type is more active, handier with its feet, and easier to keep fit.

Southerners prefer even lighter horses and consider a 1000 to 1400 pound horse to be sufficiently large. Mules vary in weight all the way from the 12-hand, 600-pound pit-mining mule to the 17-hand, 1600-pound draft mule. Large mules are usually worth more money than the smaller ones.

Head and Neck

The head is a good index of the animal's quality and should be clean cut, tapering gradually from the eyes to the poll and from the eyes to the nostrils. It should be neither large and coarse nor small and ultrarefined, but should closely approximate the length of the animal's shoulders. The eyes should be large, set wide apart, clear and quiet.

Horses' ears should be of medium size and texture and should be carried at an alert angle. Mules' ears, while larger than the horse's should be relatively trim and fine. The union of the head and neck, known as the "throat latch," should be clean-cut or free from thickness and meatiness. The jaws should be widespread to prevent choking as the animal flexes his neck while pulling.

The neck should be tapering, supple, and of moderate length. Short, stubby necks usually are found on horses that are hard to guide and hard to fit with a collar. Mules are inclined toward "ewe" necks.

Shoulders

The shoulders should be well laid-in at their top, blending smoothly into the withers. The shoulder should have approximately a 45- degree slope and a clean-cut, well-defined collar bed. Steep shoulders mean a low carriage of head and neck and consequently a short, stubby stride. They are also more apt to become blemished under hard work.

Body

The draft animal should be short and heavily muscled in his neck and loin, deep and broad in his chest, well sprung in his rib so as to have a deep, round middle, and he should be deep or well let-down in his hind flank. These characteristics insure ample lung and stomach capacity as well as a strong top for heavy pulls. Light-middled horses and mules are hard keepers. Animals with sharp backs and narrow lions lack the necessary muscling to stand up under heavy work.

Rear Quarters

The hindquarters of the draft animal generate most of the pulling power. They should therefore be wide, deep, and heavily muscled. The croup should be long, broad, and gently sloping. The thighs should be wide, deep, and heavily muscled. The gaskins and stifle should show heavy, bulging muscling. The mule is most commonly deficient in the rear quarters.

Feet, Legs, and Action

The proper placement of feet and legs has already been discussed. The feet should be large, round, sloping from front to rear, and wide at the heels. The horn of the hoof should be dense, smooth, and free from cracks. The rear feet are not as side at the heel as are the front feet. While most mules have relatively small, narrow feet, larger feet with quality are preferred.

The pasterns should slope at approximately a 45-degree angle and have sufficient length and elasticity to absorb the shock and to permit an easy gait. Short, upright pasterns go with a jerky gait and unsoundnesses, especially of the front pasterns and legs.

The cannon bones are the best place to observe quality of bone. The bone should be of medium size, dense, and flat. Knees should be deep from front to rear, wide as viewed from the front, and tapering gradually into the leg.

The hock is of outstanding importance since this is the lever used to transmit the power of the rear quarters in propelling the animal forward. The hock should be wide from front to rear, deep from top to bottom, and wide at the front. It should be clean-cut and free from blemishes and unsoundnesses.

The draft animal should have a long, straight, snappy yet easy stride. He should lift his feet high enough to clear obstacles, but not so high as to be wasteful of energy. Throwing the feet outward as they are moved forward is called "paddling". Exaggerated paddling in high-going horses is called "winging". "Interfering" is striking one foot against the other as it is moved forward. Moving forelegs around each other and planting them in a single, straight line is called "winding" or "rope-walking." Striking a front shoe or foot with a rear foot is called "forging."

"Rolling" is shifting of weight from side to side to keep balance—a defect found especially in excessively wide-chested horses. "Pounding" is hitting the ground too hard, whereas a quick, high, but short stride is described as being "trappy." Traveling wide at the hocks is a serious fault, since an animal so affected springs out at the hocks under heavy pull and punches himself in the belly with his stifle at the trot.

Common Blemishes and Unsoundnesses

A *blemish* is an imperfection or injury not serious enough to interfere with the serviceability of the animal. An *unsoundness* is an imperfection of form or function that is caused by inheritance, injury, overwork, or misuse and that interfere with the serviceability or usefulness of the animal in a way that greatly lowers his value. Work stock are usually sold with or without guarantees as to soundness as follows:

1. "Sound"—guaranteed free from any unsoundness or serious blemishes.
2. "Serviceably sound"—showing only blemishes or unsoundnesses that do not affect his usefulness.
3. "At the halter"—no guarantee as to soundness, disposition, or any other characteristic.

Unsoundnesses vary as to their effect on the usefulness of a horse or mule. It is usually accepted that the following constitute show-ring disqualifications: lameness, blindness, bone spavin, stifled, stringhalt, ringbone, one testicle or abnormal testicles in a stallion, and bad wind.

The following defects are recognized as ground for more or less serious discrimination: sidebone, bog spavin, thoropin, capped hock, cocked ankle, poll evil, filled hock, curb, fistula, umbilical rupture, and stocked ankles.

Definitions of the more common blemishes and unsoundnesses are given:

Bone spavin—a bony growth usually found on the inside lower part of the hock joint.

Bog spavin—a soft swelling located at the front inside area of the hock.

Thoropin—a soft swelling similar to a bog spavin but located in the depression just in front of the point of the hock. Pressure on one side will cause the swelling to bulge out on the other side.

Curb—a swelling at the rear of and toward the bottom of the hock.

Ringbone—a bony growth around part or all of a pastern.

Sidebone—a prominence and hardness at the hoofhead caused by ossification of the lateral cartilage.

Splint—a bony growth lying along the inside of the front cannon bone below the knee.

Stifled—dislocation of the patella of the stifle joint; usually incurable and crippling.

Stringhalt—a nervous disease resulting in a jerky, abnormally-high action of the hocks.

Fistula—a deep seated abscess at the withers.

Poll evil—similar to fistula but located at the poll.

Heaves—a respiratory disease characterized by wheezy and jerky breathing and a short hollow cough.

Blindness—a result of injury, infection, or "moon blindness" causing loss of sight in one or both eyes.

Capped hock—a prominence at the point of the hock usually caused by bruising.

Cocked ankle—usually found on rear legs with unusually steep pasterns caused by shortening or inflammation of the tendons and resulting in a short, stubby and frequent stumbling.

MARKET CLASSES AND GRADES OF HORSES AND MULES

There are no up-to-date official market classes and grades of horses and mules. They vary from market to market and from year to year. The following represents an average classification for the country as a whole.

Draft Horses

Draft horses surpass the other classes of work animals in conformation, weight, and quality. They are deep-bodied, short-legged, thick, heavily-muscled, and of good quality.

Chunks

These are similar to draft horses, having approximately the same conformation and quality, but less size. They are used principally for farm work.

Wagon Horses

These are similar to chunks in size, but are somewhat more leggy and possess more quality and speed. They are used principally in cities on milk, bakery, and other delivery wagons.

Driving Horses

The demand for driving horses has been practically eliminated by motor vehicles. The remaining demand is principally for pleasure and show purposes. Both outlets expect and will pay for quality. Roadsters and principally of Standardbred breeding and possess quality and speed. Heavy harness horses are principally of American Saddle Horse or Hackney breeding. They are a somewhat more stylish horse than the roadster and possess higher but shorter action. They are, therefore, not so fast as the roadster.

Saddle Horses

These vary in quality from the western bronco to the finest of 5-gaited saddle horses, and in size from the upper pony limit of 14.2 hands to the large, upstanding hunter of 17.2 or better. Light hunters and jumpers are usually of Thoroughbred breeding, whereas heavy hunters and jumpers often carry a cross of Percheron blood.

Ponies

This class arbitrarily includes anything under 14.2 hands in height. They vary from the chunky, rather plain-legged children's pet, to the clean-cut, stylish show pony.

CLASSES AND GRADES OF MULES

Mules are classified according to their use and graded according to their quality and value. A short description of each class follows:

Draft mules are the largest of all mules and are used for teaming, contracting jobs, and lumbering. These should be of drafty type, good quality, and straight and snappy action.

Farm mules are usually of somewhat poorer type, quality, and finish than are draft mules. They are purchased for use as farm work stock principally in the central and southern states. Mare mules are preferred.

Sugar mules are used on southern sugar plantations. They are between the draft and cotton mules in size, type, and quality. Mare mules are preferred because of their higher quality and more active dispositions.

Cotton mules are purchased for use on the cotton plantations of the South. They are definitely lighter and less drafty in type than the sugar mule. The present trends in southern farming call for larger and better mules than were used a number of years ago.

Mining mules vary greatly in size and weight according to the type of mine work they are to do. They should be of compact, drafty type. Geldings are preferred because of their more sluggish dispositions. Mules are well adapted to use in mines because they are quieter than ponies and less apt to be injured by misuse.

FACTORS AFFECTING MARKETING VALUE OF HORSES AND MULES

Class. The market class for which an animal can qualify will determine to a large extent his value. It can be seen that since 1935 saddle horses of even medium quality have outsold the average draft horse, wagon horse, or chunk.

Age. The value of a horse or mule also depends on its age since they may be either too young to work or too old to have much serviceable life left.

Sex. Mares are preferred to geldings wherever the buyer plans to raise foals for replacements or for sale. Mare mules are also preferred, except in mines, because of their higher quality and more alert, active dispositions.

Conformation. The conformation of a horse or mule greatly affects its serviceability and therefore its value. Narrow, upstanding, light-muscled draft horses sell at a considerable discount. So do low-headed, coarse, stubby saddle horses.

Quality. Quality of bone and feet is especially important since it is an indication of durability. General quality is closely related to price.

Colour. Preferences differ, but generally speaking the solid and darker colours are preferred. In teams the colours should be matched.

Temperament. Any evidence of vices or skittishness detracts from the value of an animal. However, a spirited and energetic animal of tractable disposition is preferred.

Training. The average buyer wants a well-trained animal.

Condition. Fat covers a multitude of defects. It also indicates that the animal is healthy and has been well cared for.

Style. This is especially important in saddle horses. A little work in grooming and trimming often pays big dividends.

Action. The action of a horse or mule is closely related to its serviceability. Those with short, low, sluggish, or crooked action can do less work, give a poor ride, and often become unsound due to physical defects that cause undue strain on joints or bones as well as poor action while so doing.

Soundness. An unsound or even a blemished animal should sell at a considerable discount since most unsoundnesses affect an animal's usefulness in a greater or lesser degree.

13

PIGS

Domestication of mammals has been an old and profitable business of man. It has been recorded in prehistoric period also. The domesticated mammals, in olden days, were used either for milk, traveling or for meat. There is a long list of animals domesticated by the man. These mammals help the man in various ways. Some of the common mammals which are domesticated by the man are cow, ox, deer, dogs, cats, buffaloes, elephants, horses, donkeys and pigs.

Today a number of industries are base on animals and their by-products. Thus these animals are providing handsome jobs to unemployed educated youngmen. Sheeps, goats ans swines are more successful in acilmatising themselves in each and every environment so the industries based on these animals are always in profit and they are preferred by a large number of people.

The rearing and development of sheep, goat and swine requires less economies and technology whereas the rate of production is high and duration of maturity is much less as compared with another mammal.

Rearing, development and breeding of swine is known as *piggery*. It is one of the most common and desired industry in developed countries and is gaining momentum in developing countries like India.

Pig rearing has never been a new thing for Indian people, but its raising and production has always been in primitive stage.

There is no scientific planning and investment of money only because of the reason that it is in the hands of poor and uneducated persons.

The higher class of people, educated students and economically sound parties did not show their interest in formation of pig houses and development of a piggery.

The Country Pig

The common Indian pigs are most neglected, scrub animals without any marked characteristic features. They are not paid proper attention and food for their development hence they do not get a fair chance to grow into economical animals and compete with them. These swines are small, black and produce small litters.

The meat is also not up to the mark and quite often they are infested with a variety of worms and diseases. They grow slower and die soon. The common Indian pigs have higher fat contents so they are used for medicinal purposes and the flesh is not tasty.

Advantages of pig Production

1. Pig forming requires less money, space and technology.
2. The pigs are developed in comparatively shorter period and with minimum expenditure. They pay the return in the shortest possible time as compared with any other form animals.
3. The housing and equipments are not costly
4. They do not require much labour and attention as compared with any other form animal.
5. The retrun of money is always in a handsome amount as compared with dairy and poultry.
6. The pigs grow faster than any other animal.

Efficiency in Park Production

The profit in production of the pig depends upon the efficiency of the grower in production of the pig and its marketing. He must have a look over the rate of growth, amount of feed eaten and quality of grains used in the feed.

The supplementation of vitamins, minerals, antibiotics and proteins in the feed helps in developing the swine faster and such as swine always fetch a good amount of money.

Since the hogs and swines are subjected to a large number of parasites and diseases proper case is to be taken against these diseases.

The swine used in the piggery should be of an improved quality and she must have a good breeding capacity. At the same time marketing of the pig also plays an important role in its economics.

Selection of Breeds

In establishing a piggery the selection of breed plays an important role. It depends on the will of the grower as what type of pig he wants to grow. The profit and loss depends mostly on the selection of the pig. Some of the common breeds aclimatised in India are given below. These breed can be divided into two major groups.

1. The English Class
2. The American Class

The English Class

This class of breed generally includes the pig of English origin. They are well aclimatised in Indian environment and are used for improving the race. Some of the common English breeds are :

(a) *The Large White Yorkshire*. It is an English breed but well established in U.S.A., Canada, India, Ireland and countries of Indian sub-continents. They are used for upgrading the Indian breed. These are white, big in size long and bulky with a long thin and forwardly inclined ears. They measure 350-400 kg.

(b) *Middle White Yorkshire.* A English cross breed boar developed by crossing the white large with Yorkshire extraction. It is a high class bred famous for good quality of meat. It has an early maturing and good recovery. It converts the food into meat in large ratio. These are white, large sized, with a long back and level. They measure about 270 to 380 kg.

Berkshire

This is the oldest breed of swine now established in Australia, Newzeland and Berkshire. It is famous for high yield and is used in upgrading the Indian breed. These can be recognised by its black colour and white patches on the feet, head and tail. The head is short and face is blunt. They weigh 280-300 kg.

American Breed

The American breeds are equally good for upgrading the Indian races and meat production as well. They can be aclimatised easily in Indian environment and are widely used in all parts of India. Some of the common breeds are:

Chester White

A native of Pensylvania U.S.A. is a cross breed of English white Yorkshire with Cheshire and Lincohnshire breeds. These are white in colour and produce big litters. The litters develop fast and reach maturity very soon they measure about 400 kg.

Duroc

Red coloured with white shade in colour. The colour may vary from golden to dark red. They develop fast and produce big litters.

They have high milking capacity. These are used in cross breeding weigh about 300 kg.

Hampshire

These are black pigs with white best encircring the body and fornt legs. Head and tail are black, ears straight and erected. Legs are short and produce good quantity of milk and meat as well.

Besides these a variety of porks are found which are equally good in amount of meat and its quality but require more attention. These breeds are Poland, China, Tamworth and Wessex saddleback etc. Day by day improved varieties are being produced which mature early and give flavoured quality of meat in large amount.

FEEDING AND MANAGEMENT OF THE HERD

The profit or loss of a pigger is determined on the basis of the total net gain from the piggery. This gain is calculated on the basis of the number of pits marketed. The number of pig produced in the herd depends on a variety of factors and feeding is one of them. Food feeding of the male boar and the sow is solely responsible for producing a big litter.

A protein rich diet keeps the male active and smart. It produces a good quantity of sperm with a large number of viable motile sperms, in the same way a healthy female produces a large number of eggs and they are fertilized perfectly. Such a sow produces healthy litter with maximum viability and minimum mortality. Such a sow produces less number of abnormal pigs. The diet of a lactating female is little different with the same given to the male.

The pigs have a very simple digestive system without much modifications and complications, so they require a good and protein rich diet. They are unable to digest more fibrous food but fibrous food up to a limited extent is also necessary for their proper development. A protein rich food helps the female in producing more and more milk and this effects the health of the body sow up to great extent.

A healthy baby sow changes the food quickly and starts feeding on grains and fodder where as a ill fed baby sow continues on mother milk for a longer period.

The feed of a sow should have low fibre contents such as cereal grains and their byproducts. It should have proper proportion of minerals, vitamins, medicines, and antibiotics along with the protein. It is to be in a measured quantity, which can keep a pig fast growing, healthy and strong but at the same time should not be much high in fat contents because enough of fat in a sow makes the meat of low quality. There should be a limited ratio of fat and flesh. The balanced diet of a swine comprises of the following ingradients :

Maize, ground barley or grain jowar	=	65-75%
Ground wheat or oat	=	15%
Meat scrap	=	5-7%
Soya meal	=	6-10%
Minerals	=	5-1%
Salt	=	0.5%

If such a balanced diet is given to a male swine he can be ready and it for service after 8 months whereas a female gets matured in 9 months. It is not advised to use a male in early stage, for better and fruitful results he should be used after one year and that too not daily.

He must be given rest at least for 3 days before his use and in the same way the female should not be allowed to male just after her first cycle, she must be spared for 2-3 cycles, i.e., 2-3 months after her first cycle. This cause full development of the uterus and ova and gives better result in days to come. The maturation of male and female depends upon the quality and quantity of the food, they are fed with.

The food of a meat quality of sow varies much from the food of a pregnant or milking sow they require more protein diet as compared with any other normal sow.

Gestation Period

The sow have a heat period of about 40-65 hrs. and the best period of their mating and population is the second half period during the heat period. A pig can easily be recognised during this stage because of her actions. She may mount other swines in the herd and will remain rest less and will not eat.

Her appetite is lost. She can be recongnised by examination her vulva which becomes enlarged and congested but this can be done by expert breeders only. The cycle repeats after every third week, otherwise they have gestation period of 114 days on an average which may range from 112-115 days.

Artificial Insemination

Artificial insemination in swines plays a vital role in the development of piggery. In such a condition the boar of a good health and breed is allowed to mate a dummy female with pulsating vagina and the semen is collected in the glass vials. A boar is given at least three days rest after first ejaculation so he can serve only twice a week.

The amount of sperms produce varies from 200-250 ml per boar which is kept in freezed condition after diluting with 6% glucose soln. The viability of sperms is 100% for the first 24 hrs after that the viability declines and the chance of fertilizing an egg is reduced to only 50% if such a frozen semen is used after three days.

For good litter production a sow is crossed twice after an interval of 8 hrs. This gives good result as the chance of fertilizing an egg is doubled.

In Indian artificial inseminations in swine is not done on large scale because the number of piggeries are not much and if some of them are interested in having a piggery farming they prefer to have a boar also. This ceases the scope of A_1 in sows.

Here are some practical problems associated with the artificial insemination of the swine such s the sperms have high mortality and poor ability of frozen sperms. The sow have small vagina which does not exhibit the visual symptoms of heat and they too show poor heat detections.

The sows in a herd have different days of heat so they can be managed by the boar.

The scientists in England have developed a new chemical which is fed to the female swine regularly and till the date it is fed to the sow she does not come into heat and the day the dose is checked the sow comes in heat. This has become a boon for piggery farming.

The breeders check the heat period chemically and allow all the females of the herd to come in heat on same day. This helps in A_1 and all the females give birth to the hogs on the calculated day. This provides the breeder with a benefit of arranging and managing the females of same category.

A particular type of bed. Feed, heat, water, mineral, vitamins and aids is given to the female during her pregnancy. This saves a breeder from arranging various types of beds, feeds, aids and at the same time large number of baby pigs are obtained which are marketed in a lot and the profit is obtained in a handsome amount. These things are still needed in India to improve the piggery farming.

The baby pig is given a good nutritive diet to develop fast and attain good size. Fatty diet is given only at certain intervals and after recommendation of the doctors. The well developed hog is now send to slaughter house or market. The slaughter of a pig is done after starving the pig for 24 hrs. This gives a good and standard product.

Growth of Young

For the good growth of young ones, supplementary feed should be given after 25 day by creep feeding. It should be kept in mind that sows fed on synthetic diet may not be allowed to become much fatty otherwise the size of the litter would be smaller.

The good growth in the early developmental stages plays very significant role for the development in length, muscle, weight of the pits, so up to the age of 25 days young pigs should be fed on highly nutritive and easily digestible diet.

The young pigs, with the weight of 50 pounds feeding on less nutritive diet, have good ability of fast growth to attain the maximum weight provided a good nutritive diet is given to them.

Proportional Growth of the Body

The proportional growth of the body of pigs varies with the age i.e.,the head and the legs are larger in comparison to the body but the age, the body grows very fast as a result the muscles of thigh become concave.

Such growth is found more in middle white pig in comparison to large white pig. It is very useful to have an idea about the change in the weight of the developing pigs and a proper control on that change.

The growth in pigs runs from the cranium towards backwards and from the tail forwards which meet in the lumber vertebrae as a result the loin region grows very fast followed by the pelvis and the thorax than the neck.

The head grows very slowly, it is advisable that the pigs in early stages should be allowed to fed on poor diet due to which early developing parts, as head and legs would grow very slowly and if later given highly nutritive diet the parts which grow later i.e., loin would grow very fast and the carcass with high proportion of fat is produced. In the young pigs the fat is soft and in the older ones it becomes firmer when growth rate is very fast and fat is synthesized from the carbohydrates.

Slaughter of Pigs

Prior to the killing, the pigs should be kept on starvation for 24 hours and given complete rest to get standard product. The slaughtering of pig is carried out by making it unconscious by a hard stroke given on the head, and piercing a double edged knife in the neck to shed the blood out of its jugular veins. Now carcass is washed, cleaned well in hot water and cut open to separate the various parts for different preparations.

PRODUCTS OF PIGGERY

Pork

The flesh of pig is known as pork in general and the fleshes obtained from different parts of the body have been given different name e.g., Bacon obtained from the back and sides and Ham from the back of the thigh. According to the data produced by the Directorate of Marketing and inspection, Government of India the production of pork is 5 percent of total production of meat in India.

Bristles

The wiry and stiff haris of the pigs, hogs and wild boars are known as bristles and they are obtained from the back and neck. Short bristles are obtained from the flanks and belly of the animal. The bristles obtained from the living pigs are superior in quality than those of slaughtered ones. The rough and coars bristles are generally used for varnish work and painting brushes.

The main bristle producing areas are Uttar Pradesh, Madhya Pradesh, and Punjab. The principal trading centres are Kanpur and Jabalpur but 70 percent of the total bristle export is being made from Kanpur.

Two types of bristles viz., *Desi bristles* obtained from pigs of Uttar Pradesh and Madhya Pradesh, and *Darjeeling bristles*, obtained from Himalayan foot hill and Darjeeling district. The bristles of three colours are usually found viz., white, grey and black.

The important bristle markets in the country are at Agra, Allahabad, Azamgarh, Faizabad, Jaunpur, Amaraoti, Nagpur, Santhal Parganas, Caluctta and Kakinada.

Mumbai is the biggest port to export the bristles worth millions of dollars to U.S.A., U.K., Germany and Japan. Indian Standard Institute, New Delhi standardizes the bristles and provides *agmark*. Only the properly graded bristles under 'Bristle Grading and Marketing (amendment) Rules-1962' are permitted for export purposes.

Sausages

Sausages are prepared by fresh minced pork, free from bone and skin. For the preparation of sausages usually shoulder piece, ham and bacon are preferred. The sausages are prepared by washing, cooking and mixing the pork with spices such as white pepper and paprika to make it delicious.

Lard

The fat of pig, squeezed from the body tissue is termed as lard. For the preparation of lard, carcasses are cut into pieces, minced, boiled over a furnace and the fat is removed-carefully when it is boiling hot.

Lard is used as a fine cooking medium and in the manufacture of soaps, lubricants, greases, candles and water proof materials.

Other Uses

The most important pharmaceuticals derived from thyroid, pituitary and pancereas of pigs are pepsin, thyroxin, pitutrin, insulin, liver extract, testesteron etc. The pig toe-nails are used for making tobacco fertilizer, plaster and other plastic materials. The blood is used as food for livestock and poultry and also as a manure.

DISEASES AND CONTROL

The swine production suffer heavy loss due to diseases and parasites. These parasite kill the baby pigs and check proper growth and development of the pig as adult. They damage about 60-65% of the total pigs produce, either directly or indirectly. Their control is a major factor in swine production.

Most of the diseases can be controlled by applying preventive measures and proper sanitation. Over crowding, mismanagement, roaming of pigs in open places should be checked. Vaccination and medication at regular intervals helps in preventing the diseases.

In India swines are subjected to various diseases like, plague, flue, fever, pox, foot and mouth diseases, cholera, navel ill and dysentery and tuberculosis etc. These diseases are contageous diseases and can be checked by proper sanitation and care.

The infected pig becomes weak, looses its weight, the loss in appetite and vigour is well marked. He may sit or stand alone in an unusual manner. The colour of urine and faecal matter changes and some times it is supplemented with mucous and blood also. The colour may be red, green or black.

Coughing and eye and nasal discharge is well marked in cold and flue cases. They stop feeding and go for water quite often. Parasites also play a vital role in swine business.

The common parasites are round worm, hook worm, type worms, lungworm and mange. Ticks and mites, flea, mosquitoes, lice and bugs are common ectoparasites on pigs which may cause skin disease and itch.

These worms and ecto and endo parasites are reduced in number if proper care is taken in the management of a pig house, ventilation, sanitation and space have their role in minimising the parasites. These houses should be cleaned 2 sprayed with DDT, BHC or Gamexene.

14

POULTRY

One of the outstanding features of poultry husbandry in tropical countries is that in few places is it a specialized industry and rarely does it form the sole means of livelihood or even a major source of income. Many tribes and peoples consider poultry husbandry degrading or, at best, not worth the attention of men of standing in the community.

Eggs as a form of food are taboo to some African tribes, and eggs as well as poultry flesh are forbidden to strictly vegetarian peoples. Despite this, practically every family, settled or nomadic, own some form of poultry in varying numbers so that, in the aggregate, the total poultry population in most tropical countries is impressively large. Although the contribution of poultry to human nutrition in the tropics is already appreciable, it is but a small fraction of what it could be if optimum human requirements could be satisfied.

Under present conditions the greatest incentive to the domestic poultry owner is the fact that the birds find their own food and accommodation almost without any expense to him and that everything they produce from faecal excretion to eggs and flesh earn. him a net profit.

It is largely because he is so accustomed to consider poultry keeping from this aspect and because he is so well aware of the devastation caused by oft-recurring contagious diseases that the idea of getting his livelihood by it or of making a business of it, both of which would involve financial outlay, often appears to him irrational and uncongenial.

So it would be if the enterprise was attempted in the conditions prevailing in the majority of villages or, in fact, in any place where a market was not assured for his produce at a remunerative rate.

It is obviously unrealistic to lay out capital on better stock, and food and housing for it, when it has to enter into competition with indigenous poultry which, without that expense, can be produced in quantities which meet existing demands.

Demand rises above the incidental level only where the standard of living has reached the state where animal protein is an article of diet for which provision is normally made in the family budget and where the inability of the consumer to produce the material himself necessitates his purchase of it.

There are, therefore, two conditions under which poultry are maintained in the tropics and under which production can be improved and increased, namely, that pertaining to the supply of local country family and village needs, and that for town workers and other concentrations of money-earning peoples.

Each demands a different outlook and different methods. Urban populations are rapidly growing throughout the tropics and their need for poultry produce increases correspondingly.

As long as poultry-keeping is confined to the scavenging level, it raises no conflict in priorities but, at that level, production is obviously limited in quantity: if it is increased, a regular supply of nutritious food must be provided and the interests of the domestic fowl come into competition with those of immediate human needs.

Nevertheless, as a converter of vegetable feed into animal protein, the specialized domestic hen's efficiency is high; in fact, it is estimated that, in terms of food conversion, poultry eggs rank with cow's milk in being the most economically produced animal protein, and that poultry flesh ranks above that of other domestic animals in this respect. Of the four common species of domestic poultry, namely, fowls, ducks, geese and turkeys, the first named are by far the most important in the tropics, as they are elsewhere.

FOWLS

It is believed that the modern fow s as probably originated from four wild species, *Gallus gallus* (Red Jungle Fowl), *G. lafayetter* (Ceylon Jungle Fowl), *G. sonnerati* (Grey Jungle Fowl) and *G. varius* (Java Jungle Fowl).

That its origin is tropical, accounts for the comparative ease with which any modern breed can be transferred or introduced to any tropical country, but the fowl's power of adaptation is exceptional, and its distribution is world-wide.

It has been reared in tropical Asia from time immemorial and it is therefore not astonishing that innumerable breeds and types have been evolved in all tropical Asian countries and that the same types are found in Arabia, Africa and in the East Indies in little-altered form.

In these countries the main function of the domestic fowl was to supply meat and sport and so body-size and vigour were the characters mostly sought for and cultivated. The large variety of Asiatic fowl resulted, of which the Indian *Asil* is typical. These birds are slow growers and, under the best conditions, take about six months to reach four pounds (1.8 kg.) body weight, but the cocks grow on to attain a weight of about 10 pounds (4.5kg.), while the hens average four to six pounds (1.8 to 2.7 kg.).

The breast and thighs of the cock are particularly well developed and, even in the pullets, the percentage of carcase flesh exceeds that found in less specialized breeds and makes them valuable meat producers. The average egg yield is only 35 per annum.

The egg-laying types of poultry received less attention and remain in size and productivity more like their wild ancestors but there are certain breeds noted for the relatively large number of eggs produced, although usually the eggs are small.

The *White Chittagong* of India, laying on the average 130 small eggs per annum, is exceptionally well developed in this respect but typifies the best-laying type in tropical Asia. The *Canton* in Malaysia is very similar: under farm conditions, an average of 120 eggs weighing one-and-a-half ounces (42 gm.) each can be expected, although the village average is only 80.

The cocks weigh about five pounds (2.3 kg.) and make good, if small, table birds. Other Asiatic breeds have shown a similar egglaying potential when reared on well managed farms, and even under village conditions there are breeds from which 100 eggs a hen each season are expected, but, like those of the Chittagong, the eggs are very small by European standards.

The average, tropical, all-purpose fowl is only two to four pounds (0.91 to 1.8 kg.) live body weight but carries a fair proportion of edible flesh on a compact body bearing a somewhat light covering of rather wiry feathers free from down. A naked neck is a feature of some breeds, while others have naked or nearly naked thighs.

The feather colouring is very varied but most often it is of a lighter or darker shade of brown intermingled with red or gold; black is fairly common but white is unusual: 'barring' is common. The head appendages of the cocks are relatively large but those of the hen are small. Usually some three clutches consisting of 12 to 18 eggs are laid each year, the average weight of each egg being only about one ounce (28 gm.). Broodiness is pronounced. The birds are very active, are excellent foragers and tolerate rigorous tropical climatic conditions. There are innumerable types.

In Latin America, tropical Australia or wherever Europeans have settled in tropical countries, the European breeds have been introduced and bred pure or, more commonly, used for crossing. Certain of these breeds have in the course of a generation or two adapted themselves thoroughly to the changed climatic environment and most of them maintained productivity at an economic level when the adaptation has been completely effected.

The degree of adaptation which takes place under such circumstances is well indicated by the following facts elucidated under controlled experimental conditions by Lee, *et al.*. Using unacclimatized White Leghorns, Brown Leghorns, Australorps, Rhode Island Reds and Minorcas, they found the normal rectal temperature of fowls in warm climates to be 106° F. (41.1° C.), and that it rose at a rate depending upon a rise in air temperature and, to a less extent, the relative humidity of the surroundings.

The respiratory rate corresponded fairly closely with rectal temperature. The nature of the respiration changed at 108° F. (42.2° C.) to panting with open mouth, and at 113° F. (45° C.) to deep sighing or gasping, denoting a precarious condition. The rectal temperature of all birds was little altered by humidity except at high atmospheric temperature, with the exception of the Australorps whose body temperature was raised whenever the humidity was high.

The loss of water by evaporation from the body was high when the atmospheric temperature reached 100° F. (37.8° C.) or over, or when panting occurred. The effect of air movement was slight, but a current at 300 feet (91.4 metres) a minute did reduce the rise of rectal temperature and respiratory rate in a hot, humid atmosphere but not in a dry one.

The proportion of protein in the diet did not significantly affect the reaction to a hot atmosphere. There was a smaller rise in the rectal temperature and respiratory rate and a greater evaporation from the body of hens allowed to drink *ad lib.* compared with those deprived of a continual supply of water.

When the hens were exposed to heat, with no other disturbance, a partial moult started and egg laying was reduced to about once a week, but if the exposure was frequently repeated, eggs were laid more frequently but the numbers did not return to normal for at least four months.

Eggs laid during exposure to heat were often soft shelled and misshapen. The actual act of laying in a hot atmosphere raised the body temperature two degrees Fahrenheit (1.1° C.).

In a general comparison of breed reaction it was found that in air temperatures of 85° F. (29.4° C.) no difference could be recorded, but at 105° F. (40.6° C.) with a relative humidity of 25 per cent the White Leghorns were affected least, and the Brown Leghorns more than the others.

At the same temperature with relative humidity of 75 per cent, the Brown Leghorns were least affected, the White Leghorns and Minorcas more so, and the Rhode Island Reds and Australorps were most affected. It was thought that the body conformation was the most important factor producing the difference; the long legs and exposed thighs of the White Leghorns helped to dissipate heat, while the overlapping body of the Australorps retained it.

At high air temperature it was shown that a further small rise had a greater effect on the fowls than the same degree of rise at moderate temperatures and thus any contrivance which held or reduced high air temperature in even a small degree is important.

The authors considered it probable that a dry-bulb temperature of 80° F. (26.7° C.) marks the point beyond which slight disturbances of body function in the unacclimatized birds commences. Birds acclimatized to hot climates are not unduly affected by a temperature considerably above 110° F. (43.3° C.).

From other data it appears that moderately high temperatures do not adversely affect egg weight but that exposure to extreme heat does result in a definite reduction in egg weight and that some of the weight is regained on acclimatization. The subject is fully discussed by Hutchinson.

Birds have no sweat glands and the only way in which they can lose heat by evaporation is by the lungs, hence the panting which occurs at high temperatures and the relatively small cooling effect of air currents. Well-fed poultry are less tolerant than others.

These experimental results confirm what has been learned from practical experience.

The outstanding differences between the tropical fowl and those of the improved breeds of the temperate zone are the weight of eggs laid per bird, the age of maturity for either egg or meat production, and the efficiency of feed conversion.

One hundred and fifty eggs weighing two ounces (56.7 gm.) each or some 19 pounds (8.6 kg.) in the aggregate is to be expected from hens of the improved breeds, while 70 eggs each weighing one-and-a-half ounces (42.5 gm.), or six-and-a-half pounds (2.9 kg.) in the aggregate is probably above the average of the indigenous tropical hen.

Hens of the improved laying breeds come into production at 20 weeks of age, but hens of the improved tropical breeds do not do so before 26 weeks : the specialized broilers fed on modern lines may take only eight weeks to attain a body weight of three pounds (1.4 kg.) while birds of the best tropical breeds well fed on specialized poultry farms take about 21 weeks to grow to the same size.

As there is little difference in the amount of food consumed each day, the relative inefficiency of feed conversion of the tropical breeds is apparent.

There is not so much difference between the heavy carcase producers of the best breeds of both groups, but the difference is still appreciable, being roughly four pounds (1.8 kg.) live-weight in favour of the non-tropical birds.

The only conditions under which indigenous breeds excel are those where the birds have to hatch their own brood, find their own food and maintain themselves, unaided, against predators and endemic disease.

The improved breeds of the temperate zone are of three types. The first is composed mainly of Mediterranean breeds specially developed for high egg production, of which the most popular are the *Leghorn, Ancona* and *Minorca.*

The hens of these breeds are now virtually non-broody and therefore incapable of reproducing under natural conditions. They are relatively small and are not specially grown for meat, but cockerels weighing three pounds (1.4 kg.) can be produced at four months of age.

The meat-producing breeds are typified by the *Orpington,* the *Cornish Dark* and the *Jersey Black Giant;* the average weight of the full-grown cockerel being eight-and-a-half, eight-and-a-half and eleven pounds (3.9, 3.9 and 5.0 kg.) respectively.

They are not economical egg producers under intensive systems of management and they do not tolerate high ambient temperatures so well as the birds of the other groups.

The dual purpose breeds, of which *Rhode Island Red, Australorp* and *New Hampshire Red* are the most popular in the tropics, possess many of the best features of the other types. Although these breeds are not the best for the most intensive forms of husbandry, the Rhode Island Red and the Hampshire have proved very satisfactory on tropical poultry farms: the others are not quite so adaptable but do well in selected localities.

In general it may be said that in hot, arid conditions the White Leghorns, and in areas of high rainfall with lower temperature, Rhode Island Reds have given most consistently the best results. As layers, all over the tropics, White Leghorns are the established favourites.

The breeds, listed in order of popularity, in the following countries are:

Philippines: White Leghorn, New Hampshire, Australorp, Rhode Island Red, Plymouth Rock and Jersey Black Giant.

Malaysia: Rhode Island Red, Australorp, White Leghorn, Light Sussex, Orpington and Plymouth Rock.

India: White Leghorn, Rhode Island Red, Australorp, New Hampshire, Brown Leghorn, Black Minorca and Plymouth Rock. *East Africa* (Mawali): White Leghorn, Australorp, Rhode Island Red, New Hampshire and Light Sussex.

West Africa (Nigeria): Rhode Island Red, White Leghorn, Light Sussex.

Latin America: New Hampshire Red, White Leghorn, Rhode Island Red, Plymouth Rock.

The annual average egg production on a Central Government Farm in India was:

White Leghorn	169.9
Rhode Island Red	147.1
Plymouth Rock	170.1

In Thailand in the Annual Egg Laying Test for 1952-3 the following averages were recorded:

White Leghorn	233.9
Plymouth Rock	206.4
Rhode Island Red	193.6
Australorp	139.0

On a Uganda stock farm, Trail recorded egg production calculated on a hen/day basis and on a hen housed average:

Light Sussex × Rhode Island Red	214.6	202.2

White Leghorn × Rhode Island Red	212.1	189.9
Rhode Island Red	175.1	166.2
Light Sussex	173.0	163.5
Black Australorp	167.3	153.8

The breeds mentioned above have been used extensively to upgrade flocks and populations of indigenous birds. The method usually employed is to substitute cocks of the improving breed for those of the local breed and to slaughter without exception all the latter and all the subsequent crossbred cocks until the transformation is practically complete.

The non-broody character of the high productive strains may be a handicap, as brooding hens are essential to the village industry. Crossbred hens of the specialized breeds have proved to be the most economic for commercial egg production.

Management

Exotic stock to be used to good purpose must be provided with adequate food and must be protected from weather and predators. One of five systems may be used: open range, folding, restricted range, deep litter or battery.

The *Open Range System* as practised in the temperate zone, allows great, but not unlimited, space to the birds on land where they can find an appreciable amount of food in the form of herbage, seeds and insects.

On 'clean' stubble where gleanings are plentiful up to 200 birds per acre (0.40 hectare) may be maintained, but ordinary land will carry barely half that number for any length of time. They receive the bulk of their food by hand, but the only housing provided is for night accommodation.

In the tropics, domestic fowls are almost invariably kept at liberty. Were it not that the birds are exposed to beasts of prey and to infectious disease, the system would in many circumstances be the most suitable for hens of imported breeds.

Well-fed free-range poultry seldom suffer from nutritional deficiencies, keep remarkably free from parasitic infestation and will withstand extremes of tropical weather, but because of their lethargic nature and their plumage colouring which makes them peculiarly vulnerable to marauding animals, and because of their susceptibility to disease, exotic stock cannot be maintained under this system in most tropical countries.

There is, perhaps, one breed which is an exception: the Brown Leghorn, apart from disease, is as capable of looking after itself in primitive conditions as is the fowl of any indigenous breed and, allowed fairly liberal hand-feeding, will give a very satisfactory return in eggs.

The open-range system is one which requires least expense and labour. As nature supplies so much of the protective foods to birds at liberty, what hand-feed is required does not need to be compounded with the care which must be given to the diet of the intensively managed hen. No authoritative information is available as to what is the most suitable food to supply.

The custom is to feed grain or grain offal-a custom which is convenient, comparatively inexpensive and apparently fairly effective, but the degree to which it meets requirements must depend upon the nature of the land and the foraging available to the birds.

The essentials of design in night accommodation are the same. The housing must afford comfort, adequate space, ventilation and protection against beasts of prey and extremes of weather. While most of the indigenous housing falls far short of minimum requirements, a system followed by villagers in Ceylon and described by Aitken is admirable for a few birds and might well be used in other countries where the material is at hand.

The usual size and shape of the house is that of a small bullock cart, the walls are woven leaf stalks or reeds, the roof is rainproofed with woven leaves, and the floor slatted with stalks thick enough to provide suitable roosts. The cage is hung between trees as high as convenient.

The birds have access by a bamboo pole from which the side shoots have not been removed. After the birds have gone to roost the door is closed with the pole which is then removed to the owner's house. Baffles for vermin and insects may be set on the wire or rope sling.

The *Fold System* is particularly useful for rearing chicks from eight weeks old and, under certain circumstances, for layers. It is unsuitable in arid countries during the hot weather where shade cannot be arranged. Groups of from 24 to 50 birds, according to their age, are confined in a movable wired-in run with attached housing just large enough to accommodate them at night.

The unit is moved to fresh ground at frequent intervals, preferably each day, and so most of the benefits of open range are obtained. The system allows orderly use of cropped land in a way which suits poultry and benefits the land by the even distribution of the poultry droppings. Because the flock is segregated in such small units, each individual of which is easily observed, any defect in management is soon discovered and, if disease should break out, its control is facilitated.

Complete protection can be assured against marauders both of the birds and their food, while the movable nature of the unit permits advantage to be taken of air currents and shade.

The system does involve a comparatively large amount of equipment and thus expense, and a considerable amount of labour. Constant supervision is required to see that the folds are well placed.

By systematic placing, the best use can be made of what food the land affords, such as young grass and legumes so that what is cropped one day will have sufficient time to recover before it is again used; it also prevents the concentration of parasites and gives an opportunity for their dispersion and destruction.

The control which the system affords allows quite heavy stocking of the land and, under the most favourable conditions as many as 400 pullets may be kept on an acre (0.40 hectare) but it is advisable to stock much more lightly and have a reserve of land for emergencies.

The runs are made of half-inch (1.27 cm.) gauge wire-netting on a framework of light but tough wood covering a ground surface of about 96 square feet (8.9 square metres). An easily portable house is required, preferably one that can be moved by hand over broken ground.

The slatted ark, so much used in the temperate zone, is suitable if properly modified. It should stand at least a foot (0.31 metres) above ground level. The floor is made of slats about an inch (2.54 cm.) thick placed one inch (2.54 cm.) apart.

The walls, which also form the roof, converge steeply to the ridge about four feet (1.22 metres) above the slats but, instead of meeting there, a space is left for ventilation but it is capped by ridge boards and the entry blocked by half-inch (1.27 cm.) wire-netting.

The lower two-thirds of the back of the house is usually boarded but the front, with its door, is made only of half-inch (1.27 cm.) wire-netting laid across a frame, thus permitting the free entry and circulation of air. A movable wire-netting floor is fitted about four inches (10.2 cm.) below the slats so that it stops the entry of vermin by night but can be removed during the day.

The floor space commonly allowed in Europe is:

0.25 sq. foot (0.023 sq. metre) for each 8-weeks-old bird

0.33 ” ” (0.031 ” ”) ” ” 3-months-old bird
0.50 ” ” (0.047 ” ”) ” ” 4 ” ”
0.63 ” ” (0.058 ” ”) ” ” 5 ” ”
0.75 ” ” (0.067 ” ”) ” ” pullet
1.00 ” ” (0.093 ” ”) ” ” hen.

A fifty per cent increase in floor space should be allowed in the tropics. Food hoppers and drinking vessels should be of metal. Troughs should allow 18 inches (45.8 cm.) at each side for every four hens.

Where thatching material is available, the ark can be framed with wood or light metal carrying half inch (1.27 cm.) wire-netting upon which thatch can be laid in the thickness best suited to the climate and circumstances.

The ***Restricted Range System*** is primarily used for layers. Ample housing is allowed but just sufficient outside space in pens is provided for health and comfort. Before the deep-litter and battery systems were evolved it was almost the only form of intensive husbandry practised for laying hens and where labour is dear, land restricted and expensive, disease rife and close control essential, it was found to be very suitable.

In cold climates the houses are made large enough to accommodate all the flock during periods of inclement weather, litter being spread over the floor to provide scratching and exercise, therefore fairly generous floor space was allowed, that is, about four square feet (0.37 square metre) per bird.

There is an out-run at both sides of the house, one being used as the other is rested. Each affords about 12 square yards (10 square metres) of ground per bird and thus one unit of 150 occupies slightly less than an acre (0.40 hectares) of land. Where conditions are favourable, stocking up to 200 hens or up to as many as 400 growers is practicable.

Under tropical conditions the maximum space for each bird as detailed above is adequate, but it is advisable to have units of not more than 50 birds because closer supervision is needed. The arrangement means an increased cost in housing, a little more land is required, and the day-to-day work is somewhat heavier but in most circumstances the extra expense is justified.

The wire out-runs should be at least six feet (1.8 metres) high and should be of half-inch (1.27 cm.) heavy gauge galvanised wirenetting with the topmost six inches (0.15 metre) left slack so that it does not offer a bird a steady foothold. Shade must be provided. As the housing is to be of a permanent nature it pays in the long run to have it constructed of durable material.

The essential features of the house are: that it should be orientated east and west; the floor should be impervious, and cement best fulfils requirements but well-beaten earth or clay is a good substitute; that the frame should be of light metal and the walls of cement-faced cavity brick or some such prefabricated material as corrugated asbestos or thick plastic; the use of wood should be avoided as far as possible and when used it should be dressed and protected against termites.

In dry countries the walls should be eight feet (2.44 metres) high at the eaves. That side of the house least exposed to the sun should be walled to about two feet (0.61 metre) and wire-netted above. The opposite wall should be built to four feet (1.22 metres) and the remainder netted. Light thatched shades which can be dropped over the netted spaces as a storm shield are invaluable additions.

In wet climates, driving rain may necessitate the closure of the house on all sides, in which case ventilation should be provided from drop shutters placed 18 inches (45.8 cm.) above the floor on one side and under the eaves on the other, and light from windows running the full length of the lee side of the house.

The Philippines Department of Animal Husbandry (1955) found that of four types of houses roughly eight feet (2.44 metres) square and four-and-a-quarter feet (1.30 metres) high, all constructed and roofed in the same manner and connected to out-runs by an open front, the one whose only walls took the form of a 14-inch (35.5 cm.) protection at the level of the roosts was more suitable for laying hens housed in it than the one which had no walls or the three-walled one with a gap of about a foot (30.5 cm.) running round the bottom. All were better than the one completely walled from floor to roof.

The roof should overhang the walls by at least two feet (61.0 cm.). It should be made of corrugated asbestos, aluminium sheeting, wood or corrugated iron and, when practicable, covered over with thatch.

Very satisfactory prefabricated sectional housing is now on the market.

All the fittings should be easily movable. The perches should be at least one inch (2.54 cm.) broad and need not be more than two feet (61.0 cm.) from the ground. Twelve inches (30.5 cm.) should be allowed each bird so that there need be no crowding. At least a dozen nest boxes each with a minimum floor space of one square foot (0.093 square metre) should be placed along the east end of the house about two feet (61.0 cm.) from the floor and one foot (30.5 cm.) from the wall. Communal nest boxes are suitable: a convenient size is six feet (1.83 metres) long, two feet (0.61 metre) broad and 16 inches (0.41 metre) high in front. Metal food hoppers and water troughs should be raised a foot (30.5 cm.) from the ground.

The *Built-up Litter System is* so called because the poultry are kept continuously indoors on a layer of litter so laid and maintained that it encourages the multiplication within it of micro-organisms which break down the droppings and, eventually, the litter into a mass of fine, dry, friable material which, except when it smells of ammonia, is practically odourless and is comparatively hygienic.

The system is extensively used in non-tropical countries both for layers and broilers, birds intensively reared up to about three pounds (1.4 kg.) bodyweight, as well as for rearing and even for breeding stock. Its chief advantages are that it allows great economy of land and labour, and that in cold climates the litter, by insulating the birds from the coldness of the surrounding ground and by itself generating a certain amount of heat tends to counter the adverse effect of wintry seasons.

As none of these considerations is relatively of much importance in the tropics, and as heat loss rather than heat conservation is desired, it might be thought that the system was unsuitable. On the contrary, it has given satisfaction in tropical climates when a properly constructed house permits a controlled environment.

Its success is based upon the ability to provide the stock with what is considered to be the optimum environment for whatever production is required. This demands complete control of ambient temperature, humidity, air movement, light and space.

The building must therefore afford effective insulation from outside atmospheric conditions, sound, sanitary flooring, forced cool air ventilation, and artificial light, as well as such mechanical apparatus as will facilitate day-to-day maintenance.

If such capital outlay as this entails is to give an adequate return, the cost must be distributed over a large enough population to bring the share per bird to a competitive level.

Other things being equal, therefore, a large unit is more economical to run than a small one, but the larger the unit the lower the level of husbandry inevitably becomes, the rate of production decreases as population density increases and the chances of introducing disease, as well as the consequences of doing so, grow with the size of the unit.

The ultimate size of the plant must therefore be closely related to the competence of the management, the skill of the stockmen and the reliability of the servicing available. Osbaldiston and Sainsbury have reviewed the accepted standards of construction which these considerations demand. Among the many detailed recommendations they record, only a few can be mentioned here.

Assuming that the house has insulating walls and roof, and a three to four inch (7.6-10.2 cm.) concrete floor underlaid with a dampproof course, the maximum capacity should be no more than 16,000 broilers or 5,000 layers with up to one square foot (9.29 sq. dm.) for each layer on deep litter. Any of the classical systems of cool air ventilation may be employed but the design should be as simple as possible and easy to control and operate.

Efficient ventilation is of the first importance. A fan capacity of up to one cubic foot (0.03 cm.) of air per minute per pound (0.45 kg.) bodyweight for broilers and twice that amount for layers is needed. The ambient temperature may range from 21° C. (70° F.) in a broiler house with chicks, to 10 to 13° C. (50-55° F.) for adults. A relative humidity of 35 to 70 per cent should be maintained.

Light should be of even intensity throughout the building, and lighting points should be no more than 15 feet (2.8 m.) apart. Continuous light reduces age at maturity but increasing day length is a greater stimulus to egg production than absolute day length. The ration of light for layers should be increased from six hours at 20 weeks old by 18 minutes each week for the following year.

Feeding space at troughs should be 0.5 inches (1.27 cm.) per bird up to three weeks old, and three inches (7.6 cm.) thereafter. Watering space should not be less than 0.2 inches (0.5 cm.) per bird.

Where a lower level of production will suffice, a smaller unit with comparatively inexpensive accommodation is all that is required. It is important that the house should be so designed and constructed that it is as cool as possible inside, that there is ample ventilation and, in damp climates, that the litter is kept dry.

There is often difficulty in providing plenty of fresh air without admitting rain or dampness, or in providing shade and protection against strong winds end sand-storms. The best design has not yet been evolved but, in general, that indicated as suitable for the semiintensive system, if a breadth of about 20 feet (6.1 metres) is allowed, can be used as a basis. The walls should be carried to a height of nine to ten feet (2.7 to 3.1 metres) so that there is always working room above the litter and that heat from the roof can be dissipated.

In dry climates thick mud walls well overhung with a heavily thatched roof makes a good, cool house. Sunlight should not be allowed to fall directly on the litter.

The material most often used for litter is peat moss, wood shavings, sawdust and chopped straw. In the tropics sawdust with or without wood shavings gives good results. Where there is a tendency to dampness, the shavings 'open out' the bed and make it easier to turn over and work. Some tropical straws are so thick and fibrous that they are not easily disintegrated and are unsuitable, even when chopped.

Tropical hay, on the other hand, when well broken or chopped makes useful litter especially if it is put on a layer of old horse dung to start the bacterial process and later mixed with sawdust. If careful attention is given to turning the bed when necessary, to keeping it evenly distributed and in the right state of dryness, almost any crop residue, such as groundnut husks, rice husks or cereal chaff and bagasse can be used.

For full-grown birds an initial layer about nine inches (22.9 cm.) deep is required. The bed should need nothing added to it for some weeks and until the litter has been broken into a dry, crumbling state when another few inches of material are put on.

More additions are made from time to time as required. If it is not thoroughly scratched by the birds or turned by hand, moulds are apt to appear on the surface, or underlying dampness is indicated by surface caking, when the bed should be again forked up and fresh litter added.

If all goes well the bed will last almost indefinitely but as a precaution against disease, especially in heavily stocked units, it should be renewed at each change of stock. The house should be left empty for a week or two, and all the fittings should be thoroughly cleaned and disinfected. It should be constantly borne in mind that if disease gains entry, all susceptible inmates will almost inevitably become infected.

Young stock should not be accommodated where adult birds have been maintained. When chicks are placed on litter, an initial depth of two or three inches (5.1 or 7.6 cm.) is sufficient and the layer can be built deeper as the birds grow.

The built-up litter can be used for adult birds or for fattening cockerels after it has been vacated by the chickens, provided they have remained healthy. When it is necessary to re-use old litter in temperate climates, it is usual to heap it for some time because it has been shown that the heat generated by fermentation inside the mass is high enough to kill many, if not all, of the harmful microorganisms.

Within the house all the fixtures should be at least 18 inches (0.46 metre) off the floor and be easily movable. A series of perches 15 inches (0.38 metre) apart should be provided to give nine inches (22.9 cm.) to each bird. It is convenient to have them placed over a pit which is covered with three-inch (7.6 cm.) mesh wire-netting through which the droppings fall.

Feeding troughs and hoppers, so constructed that birds cannot stand on the food, should be 18 inches (0.46 metre) above the litter level, with perches placed fourand-a-half inches (11.4 cm.) from the trough sides. A supply of clean water should be similarly arranged and precautions taken that overflow and splashing are prevented or run off so that the surrounding litter is not spoilt by the promotion of undesirable decomposition.

Where a favourable climate makes it suitable, a deep-litter house can be provided with an out-run so that the benefits of both systems can be obtained. In Mawali, this form of management is encouraged by the provision of a house described by the Director of Veterinary Services thus:

'The deep-litter house is a simple wood and chicken-wire structure prefabricated at the Poultry Centre and transported in sections for erection throughout the rural areas for use as a distribution subcentre. No foundations are required, the structure being placed on a single line of concrete blocks or bricks. The roof of the run, which is thatch, is put on locally after its arrival at the site.

Under skilled management the *battery system* of housing has proved to be the best in regard to egg production, efficiency of food conversion and mortality. The hens, either singly or in pairs, are confined to a cage. The usual floor space allowed is 15 × 18 inches (38 × 46 cm.) for singles and 18 × 18 inches (46 × 46 cm.) for doubles.

The height should be 17 inches (43.2 cm.). The floor, made of stout galvanized wire, slopes four inches (10.2 cm.) from behind to an egg cradle extending some six inches (15.2 cm.) in front of the cage, into which the eggs roll as they are laid. Underneath is a tray for droppings. Both food and water receptacles are outside the cage. The whole structure should be of metal so that no parasites will be harboured and so that thorough disinfection can be carried out as often as required.

Provided the batteries of cages are set up in a place which is well ventilated and lighted, is not too hot and is vermin-proof and that the food meets all nutritional needs, the system has proved to be remarkably successful in the tropics.

It may be that as it requires a minimum expenditure of energy from the bird, which spends all its time in the shade, it lessens the load of excess body heat. At any rate, at has the advantage of saving space, as the batteries are set up in

tiers, of providing the best opportunity for close individual supervision so that the capabilities and productiveness of each bird can be ascertained, of minimizing the chances of infection and helping to prevent the spread of disease and of giving each bird the quiet and effortless life which is conducive to maximum egg production.

An overhead shelter with wire-mesh sides may most suitably meet housing requirements, but insulated housing and the controlled environment described earlier are needed for maximum production.

The servicing of each cage entails much labour if it is done individually. Modem factory-produced batteries are fitted with automatic or easily worked feeding, watering and cleaning devices which, as long as they are kept in repair and are powered from a reliable source, are excellent.

Where land is cheap, labour not expensive and skilled supervision scarce, the system has not much to offer. The success of any intensive system of poultry husbandry depends upon the use of birds of specialized breeds, the supply to the birds of all their nutritional requirements and the effective protection from disease, parasites and marauders.

Eggs for Hatching

A hatching average of over 65 per cent of all eggs set is considered to be satisfactory and is seldom consistently attained in the tropics. Infertility should not be more than 16 per cent, although a level much above this is often recorded. At least 85 per cent of the eggs retained after testing should hatch if the incubator and its management are good.

Presuming that both have been efficient, failure to hatch may be because of infertility, defective nourishment of the breeding stock, damage to the egg through faulty handling or storage, physical defects of the egg, hereditary defects and disease.

Infertility attributable to the cock may be due to several causes. The bird may be too young or too old, males of the light breeds may start to mate when only ten weeks old but the full mating power is not reached till about seven to eight months of age, and vigour usually markedly declines after the age of two years.

There may be too many hens; in pedigree breeding pens the maximum number should be 12; in flock-breeding up to 15 or more may be allowed. Cocks often show preference for certain hens and neglect others and in pens with only one cock considerable loss may thus result until either the cock or the hens are changed.

Fertile eggs may be laid within 30 hours after mating but fertility is uncertain. The maximum number of fertile eggs are not laid until five days after the cock has been put with the hens. When he has been removed, fertile eggs will continue to be laid for ten days or a fortnight but after that there is a marked reduction in their number although occasional ones are laid for three weeks or longer.

If the sire of a chick is to be known with some certainty, the hen should not be mated until she has been at least ten days away from a cock.

If either hens or cocks have for any length of time been fed a diet deficient in any of the essential constituents, fertility may be diminished or the chicks may be defective and die during incubation or soon after hatching. The vitamins are of special importance. A deficiency of vitamin A adversely affects the reproductive cells.

A deficiency of certain members of the B group results in death or deformity of the embryos at different stages of development: any chicks which are hatched may be deformed or may be so weak that they have a poor chance of survival. A lack of vitamin E results in the death of the embryo about the fourth day of incubation.

Deficiency or excess of certain minerals have almost equally bad results, but it is the results of an inadequate vitamin supply which are so often seen in poultry kept under intensive systems in tropical countries.

Eggs should remain fit for hatching up to seven days with certainty provided they have been stored at or near the optimum temperature of 55° F. (12.8° C.). If they are kept in higher temperatures the fertility and hatchability decrease appreciably and, even when stored at 55° F. (12.8° C.), if the ambient temperature reaches 100° F. (37.8° C.) before they are collected, their reliability has been found to be reduced. On the other hand, opposite extremes of temperature are equally deleterious and if, for instance, the eggs be put in a refrigerator and held below freezing point for even an hour, abnormalities during subsequent incubation follow. Nevertheless it is generally not necessary to gather eggs more than once a day even in ambient temperatures of 110° F. (43.3° C.).

During storage, eggs need not be turned periodically but they should be placed with the broad ends up.

Much agitation during transit before incubation lowers hatchability but the effect is considerably reduced if the eggs are allowed to stand at rest for at least 24 hours. Air transport at the height of over 38,000 feet (11,582 metres) did not affect hatchability but vibration as well as chilling and sudden reduction of air pressure had the usual effect.

In the hot season the number of abnormally shaped eggs and those with soft shells is increased. Obviously eggs of that description should not be used for incubation nor should those which are either much larger or much smaller than the average of the flock, but eggs of approximately the same weight and age, with a good, sound, smooth shell should be chosen. Eggs which are visibly porous are particularly unsuitable in hot dry climates.

Hereditary defects are perhaps not as common as might be supposed. A number of deformities and monstrosities which were formerly considered hereditary in origin are now known to be due to other causes. Nevertheless at least 20 lethal mutations have been recorded and probably much of the malpositioning of the chick in the shell which is such a potent cause of unhatchability may be directly due to hereditary factors.

Several diseases are passed through the egg but only a few affect either fertility or hatchability. The commonest disease, or at least the best-known one, is bacillary white diarrhoea, or 'pullorum disease', which causes the death of a certain number of infected chicks in the shell and of a great number not long after hatching. Infected hens are detected by blood tests, as are those carrying other diseases and it is therefore advisable to obtain eggs for hatching from such tested birds.

Incubation

When calculating the number of eggs which must be incubated to produce a given number of laying pullets, the following facts must be borne in mind: only 50 per cent of the chicks hatched will be pullets; only 60 per cent of the eggs will probably hatch; some ten per cent of the chicks will die before attaining maturity; some ten per cent of the pullets will have to be culled.

Thus four eggs must be incubated for every laying pullet required. Except under intensive husbandry the best results are obtained when hatching takes place during the cool season in dry climates and during the cessation of rains in humid ones. At other times not only is the hatching percentage of incubated eggs lower but the chicks are handicapped by unfavourable weather.

It is by natural incubation that most chicks of indigenous breeds are produced. As yet, modern style commercial hatcheries are virtually unknown but large hatcheries equipped with artificial incubators of various sizes are a part of all large government poultry farms. Small artificial incubators, in spite of their relative inefficiency, are being increasingly used by private owners.

The age-old hatching system of the Orient and Egypt is still employed extensively in South-east Asia. The operators require great skill and experience. These incubators, like modern ones, are best situated in a room which has been more or less insulated and where air movement and humidity can be regulated with some accuracy.

The initial heat is derived from a slow fire, the sun or even from fermentation, and is continued in some cases by that generated from the growing embryos. The eggs are systematically arranged on an earthen bed, in wood trays or in baskets which are suitably insulated in rice husks, chaff, blankets, felt or such material.

Frequent placing, turning and examination of each egg necessitates constant work, but skilled operators dealing with thousands of eggs at a time obtain a hatch of 75 to 85 per cent.

The village hen, sitting on the usual clutch of some eight to ten eggs offers a striking contrast. She needs little food and less attention but, under prevailing conditions, the result is generally comparatively poor or, if the hatch is good, the mortality which follows is high.

If the incubating hen is given only a fraction of the care needed for the artificial incubator, the full benefits of the natural process are obtained. The principal requirements are: a plentiful supply of drinking water and a subsistence supply of food, protection against insect pests and vermin, and, on hatching, protection of the chicks from predators.

The nest should be placed in some quiet, cool corner at ground level and should be enclosed so that the hen can leave the nest only when permitted; that should be each morning, for about ten minutes, as soon after dawn as convenient, and again in the evening. During that time she should be given grain to eat, fresh water to drink and a sand bath in which to dust herself. She should be left undisturbed after the eggs have started to chip and allowed to move off the nest into her brooding quarters at will.

In very dry climates the nest should be damped daily. Chaff, sawdust or sand liberally mixed with a parasiticide such as sodium fluoride, makes the best bedding.

Artificial incubators can now be had to suit almost any circumstance and, if those obtained from reputable firms are operated strictly in accordance with the instructions of their makers, they perform their function admirably. When disappointing results follow their use, those not due to faults of the operator can generally be attributed to defects of the eggs, the incubating room or the source of heat.

The very small incubators are particularly sensitive to adverse surroundings and the instructions concerning the initial heating and the maintenance of proper humidity and heat need to be intelligently interpreted and meticulously observed. The larger incubators; and especially the mammoth types, carry such valuable loads that common prudence demands that they should be in the charge of trained operators.

During incubation the essential air conditions are that there should be sufficient oxygen in circulation to supply the needs of the growing embryo, that the carbon dioxide resulting from embryonic metabolism should not be allowed to accumulate, that there should be a relative humidity which allows not more than ten per cent of the water content of the eggs to be lost and that the temperature is such that life within the egg is maintained at the optimum level.

In the tropics the maintenance of humidity at the proper level usually gives more trouble than the maintenance of the right amount of heat. In small incubators about 58 per cent relative humidity is needed at 102° F. (38.9° C.) up to the eighteenth day and in the larger ones, at 100° F. (37.8° C.), 60 per cent is required but thereafter till the chick is hatched up to 70 per cent is needed.

For incubators with unassisted air circulation the usual temperature recommended is 101° F. (38.3° C.) during the first week of incubation, 102° F. (38.9° C.) during the second and a maximum of 103° F. (39.4° C.) during the third. The thermostat fitted to the machines readily makes adjustments necessitated by moderate changes in outside temperature, but when the diurnal variation is very great, the thermostat may not respond with sufficient speed to give optimum results, so it is advisable to have the incubator situated in a fairly well insulated room.

Such a room by itself does not help much to regulate humidity, for in certain climates, changes of as much as 55 per cent in relative humidity within the room have been recorded during a day, and emergency action has to be taken to supplement that of the moistening mechanism with which most modern incubators are provided.

If the oxygen and carbon dioxide level is to be effectively maintained there must be a passage of fresh air not only over the eggs but the exchange must take place around the eggs and therefore all air pockets must be eliminated. This is facilitated by placing the incubator away from the walls and corners of the rooms so that there is space for ventilation and for work.

Incubating eggs should be turned at least five times daily up to the eighteenth day. All modern incubators provide means of doing this easily. If turning has to be done by hand, less than five times must often suffice, but it must be done not less than thrice.

At the seventh and sixteenth day of incubation it is customary to check and remove all infertile eggs and those containing dead embryos. This is done by 'candling' which consists of passing a beam of light through the eggs when they are placed over a slit or diaphragm which excludes surrounding rays, so that the state of the egg can be seen with ease.

Infertile eggs show up as 'clears', dead 'germs' are seen as a dark spot of varying size and degree of density, live ones cast a shrimp-like shadow from which blood vessels radiate, dead embryos do not make a uniform darkness but appear to have an air space at both ends of the egg, hatchable eggs are uniformly dark with a clear-cut air space at the broad end of the egg.

The hatched chicks should be transferred from the incubator to the special trays provided.

Brooding

The newly hatched chick during the last stage of incubation, has absorbed with the egg yoke all its nutritional requirements for at least the first 48 hours of life but it has been demonstrated under tropical conditions that it pays to make food available soon after hatching. Attention must, in any event, be given to the bird's surroundings.

If it is left in the hatching tray for the first 24 hours, the relative humidity must not be allowed to rise above 55 per cent because high humidity with high atmospheric temperature has a most debilitating effect.

If the chicks have been hatched under a hen, she should be left undisturbed for two days. If they have been artificially incubated, they should be transferred to a brooder as soon as convenient after they have dried out.

There are many designs of brooders but all allow for accommodation at a temperature where body heat is steadily maintained, and floor space where, at a lower temperature, the birds can exercise, feed and drink.

If the brooder has been designed for use in cold climates, as so many are, it is advisable to reduce the number of chicks held in it at any one time by at least one-third of the number recommended by the maker: 75 square feet (6.97 sq. m.) per 100 birds should be the minimum floor space allowed.

As a precaution against parasites and disease, the floor should be of wire-netting so that droppings pass through out of harm's way. This arrangement is very convenient in hot climates for it helps to lower the temperature of the feeding area and, when necessary, the inner accommodation as well.

The initial temperature in the brooder should be 95°F. (35°C.) two inches (5 cm.) from the floor and should be reduced by 5° F. (2.8°C.) each week till it reaches 70°F. (21°C.) which, at this stage of growth, is the optimum temperature outside the brooder.

In the tropics artificial heat may be necessary only during the night to recompense for the drop in temperature. Where electricity is available, low-hanging infra-red lamps are most satisfactory but ordinary electric lamps or kerosene ones provided with a reflector to throw the heat down into a curtained area do very well.

Accommodation under any system must be adequately ventilated. Drinking water must always be at hand, that is, within six feet (1.6 m.) of young chicks.

In most cases '*cold brooders*' prove adequate, especially when the number of chicks is not very large. Such brooders merely provide sheltered space in an insulated box-like chamber usually under strips of cloth, or some such material, hung so as to allow easy entrance and exit for the chicks. The warmth required arises from the bodies of the chicks and is conserved by the chamber. About 12 square inches (77.4 sq. cm.) of floor space is allowed for each bird.

Sexing. It is of economic importance that the sex of the chick should be known as early as possible. The Japanese method of 'sexing' day-old chicks by delicate manipulation of the cloaca demands experience and skill which only fully trained experts possess. The best of these can guarantee results with an error of not more than five per cent. Optical sexing instruments are now available which can be used satisfactorily after comparatively little practice.

A more useful way for commercial breeders is to take advantage of hereditary feathering characters which are sex-linked. Newly hatched Brown Leghorn chicks, for instance, are marked differently according to their sex, for pullets have a broad, brown strip, clearly seen, running from the comb down the back of the neck, while in cocks the strip is narrow with borders and end not clearly defined.

A more striking difference occurs when cocks bearing golden plumage are mated with hens which have a silver-like, but not white, plumage, such, for instance, as a Rhode Island Red cock crossed with a Light Sussex hen when the male chicks on hatching have silver plumage and the female golden.

'*Barring*', that is, bands of darker colour running across each feather, which is a common feature of many tropical breeds and which is characteristic of Plymouth Rocks, is also a dominant sexlinked character, so that the chicks of barred hens and non-barred cocks can be distinguished by a difference in the shade of the down. Late feathering is another dominant character which can be used. The chicks of reciprocal crosses are not differentiated.

Rearing

The risk of death of chickens during the first few weeks of life is very great, particularly in the tropics. When reared by a hen under village conditions, the loss from predators, accidents and disease is very often over 50 per cent. When reared artificially, unsuitable quarters causing overcrowding or overheating, unsatisfactory feeding, and such common diseases as coccidiosis may take as heavy a toll.

All these losses are preventable; they do not occur under good management. Protection against marauders can easily be devised. There may be some difficulty in supplying all the nutritional needs of confined stock, but that can be assured with expert advice.

What is important, then, is that clean accommodation with ample ventilation in a dry atmosphere at a constant temperature should be provided and maintained. As the risk of disease during this period when the birds are in such close contact is so great, it is wise to take the protection offered by the use of antibiotics such as aureomycin (chlortetracycline) mixed in the food at the rate of five to ten parts per million.

When coccidiosis is to be feared, sulphaquinoxaline at the rate of 0.0125 per cent, or nitrofurazone at a level of 0.005 per cent should be added to the food. The distribution of such small amounts throughout a mass is the work of the specialist.

When the weather is very hot, if the chicks appear to lose vigour, it does not necessarily mean that they are sick but it usually indicates that more space and air is required. Quarters which appear airy and suitable when unoccupied in the cool of the morning may be unbearably hot and airless when the chicks are assembled in them. It is then that the adequacy of the accommodation should be judged.

When they are about six weeks old the chickens should be transferred to folds or deep litter where the pullets will remain till a month before they are expected to come into lay. The ones chosen for retention are then placed in their permanent quarters-in batteries, on deep litter, in pens or in folds.

If the birds are at all crowded, even when a few weeks old, featherpicking and cannibalism commonly occur. The exact cause of this unnatural habit is unknown. Its incidence can be appreciably reduced by keeping the birds in continuous light of low intensity.

If noticed early, if the first offenders, and all victims are removed and if a change to better quarters is made, this so-called vice can be stopped, otherwise some such means as cutting off about one-third of the upper beak must be practised. 'De-beaking' is effective and has no ill results when the birds are fed on mash from troughs; the beak grows again in the course of time.

Preventive vaccination against fowl pox should be undertaken when the birds are six to eight weeks old and again when they are transferred to permanent quarters. They should also be inoculated against fowl pest and fowl cholera at that time. Any bird which is observed not to be in good health should be immediately removed from the flock. It seldom pays to treat sick individuals especially when they are suffering from an infectious disease.

Production

The system of management to be used will depend very largely upon the object for which the birds have been reared-for breeding, market egg production or for meat. Under any system, only those birds which are healthy and have the appearance of good producers should be retained; all others should be culled as soon as their inferior qualities are perceived.

Hens for commercial egg production can be maintained under any of the systems described above. They should be from a proved good-laying strain. The individuals are usually selected on their physical appearance. Those of a good layer, while in lay, are: clear bright eyes; red fleshy comb and wattles in contrast to the pale, shrivelled ones of the non-layer; moist, oval shaped, large, white vent; broad spacing between the pelvic bones which are thin and pliable-they should be three fingers' breadth apart; a deep, welldeveloped abdomen, covered with soft, pliable skin.

Pullets should be placed in their permanent quarters about one month before they are due to lay. A change in quarters, even from one cage to another in the same battery, or a major change in routine, may cause a temporary fall in egg laying.

In the temperate zone, pullets of the light breeds should be in lay when six months old, and those of the heavy breeds a month later. In the tropics, even under the best conditions of management, laying tends to be delayed somewhat beyond these ages, and the indigenous pullet usually does not begin until about nine months old.

The delay in egg production is not altogether due to heredity. The ovaries will not function until body growth is practically complete and this is often delayed by inadequate feeding. When sexual maturity is reached it appears that light affects the hormone system activating the ovaries and its effect increases with the length of exposure.

As there is little difference in the length of the tropical day there is little seasonal change in egg production throughout the year unless the light ration is controlled. The pause of a day or two in the laying rhythm is an individual hereditary character.

Table poultry, in tropical countries, are nearly always by-products of market egg production; the cockerels and old stock have to be disposed of in the most economical manner possible in a market where quantity is the chief consideration of the consumers. Discards from the housewife's flock still form the main supply of the large consuming centres and have a permanent depressing influence on the price of poultry meat.

Every effort is therefore made to dispose of table birds as early as possible and to keep them only so long as they will continue to grow on inexpensive local produce such as rice and sorghum bran and other cereal offal. When large numbers are involved, deep litter is the best method of management: with smaller numbers, free range is the cheapest.

The tropical cities with their increasing population and wealth appear to afford the steady market for the output of the modern broiler house in which thousands of chicks are intensively grown to weigh some four pounds (1.8 kg.) at nine weeks old.

Breeders

Breeders should be selected on performance, physical condition and on health record. The most important characters to assess are the mean performance of the daughters of the cock or the hen or their sisters and, of these, the factors to consider are: age at coming into lay, annual egg production, weight of egg, hatchability, viability of chicks, body weight of the adults and adult viability. All these important aspects of poultry-breeding are discussed by Mann.

Having chosen the best cocks and hens, they should be maintained under conditions where the further performance of each can be kept under review. A cockerel is more prolific than a cock. A fully developed pullet makes as good a breeder as a hen.

The fold system of husbandry is suitable for pen-mating and the semi-intensive system for flock-mating. Both afford protection against climatic extremes, or make its provision easier, and they do give that degree of freedom and approximation to natural conditions which help to maintain health, vigour and fertility.

Trap-nesting may be practised, that is, each nest-box is constructed so that the hen shuts herself in it as she enters and is retained until released; in this way the productivity of each hen can be recorded and each egg marked for identification. Trap-nesting is not the only method of recording egg production. Under certain circumstances group recording may be more desirable and it is always cheaper.

It is particularly important that breeding stock are adequately nourished and not exposed to extreme heat.

Feeding

As a first food for recently hatched chicks, millet or finely broken grains of any of the cereals is suitable, especially if mixed with milk in any form. If a mash of these fortified with other foods is offered it may be observed that sizeable grain particles are selected from the mash and the rest is left.

Quickly, however, the range of selection increases and soon all the contents of the mash are consumed. From the first day, fine, hard grit should be always readily available.

Poultry differ from other farm animals in that their sense of taste and smell is poorly developed and to them the palatability of a food largely depends upon its appearance and feel.

Light-coloured material with a smooth surface appeals to them, and foods with irregular, rough surfaces which cause discomfort when being swallowed or mashes which, because they are too light and dusty or too wet and sticky, clog the nostrils or the beak, sooner or later cause aversion. Similarly, foods which distend the crop by swelling on becoming moist or which clog the gizzard because of their fibrous nature become unacceptable.

The capacity of birds of the light breeds is limited to about four ounces (113 gm.) a day, and of the heavy ones to some five ounces (142 gm.). For maximum production these amounts of easily digestible food containing all nutritional requirements must be consumed daily.

That cannot be done if the food is too bulky, which usually means that it contains too much fibre. A ready means of testing this is to measure one ounce (28 gm.) of the meal into a standard 100 ml. glass measuring cylinder and, having tapped the sides until the material has settled, to read off the volume on the scale. Meals which occupy a space of over 70 ml. are too bulky.

Whatever system of poultry husbandry is followed, it is advisable to feed a complete and properly balanced ration; in all systems of intensive husbandry, it is essential. One of the reasons for the low production of the indigenous village hen is that it gets insufficient food.

When it is fed properly, a threefold increase in production has followed. The food for poultry under extensive systems of husbandry may be compounded by following general principles which practice has proved to be satisfactory but maximum production is not likely to be attained. Under intensive husbandry something more exact is required and the specialist's knowledge of the commercial feed compounder is usually necessary.

The ability to carry out the detailed analysis required, the possession of the special plant needed for processing and the facility to buy in bulk allow the feed merchant to provide a near optimum ration without which intensive methods become uneconomical.

The amount of food a healthy bird can consume each day does not mean that, given the opportunity, that amount will be eaten. Its appetite is determined by its energy requirements, at least until it reaches maturity. It is only then that sufficient will be eaten to build up a reserve of body fat.

Till then the bird will usually eat just enough to meet its immediate requirements. Thus, if high energy grain such as sorghum, maize or rice forms the basis of the ration, less will be eaten than if it is composed of grain of low calorific value such as the bulkier barley or millet. If the ration is to be properly balanced a correct ratio must be maintained between the protein and the metabolizable energy of the food consumed, and the other necessary constituents, such as vitamins and minerals, must be included.

That is done with an exactness by the specialist which cannot be attained by the ordinary poultry keeper. The latter with the help of published lists of representative values of his feeding stuffs can compound a ration which complies with sound standards such as those listed below based on the recommendations of the Agricultural Research Council of the United Kingdom. The publication should be studied for the more detailed information of the nutrients required for poultry.

Table 14.1 : Some Nutritional requirements of poultry.

Class	*Metabolisable energy Kcal/lb. (kg.)*	*Crude protein %*	*ME/P ratio lb. (kg.)*	*Ca. %*	*P. %*
Chick starter	1340 (2950)	20	67(147)	1.0	0.7
Chick grower	1320 (2900)	16	82(181)	1.3	0.6
Layer	1430 (3150)	15	95(210)	3.5	0.7

Where the ability to select material is restricted by local and immediate considerations the owner mixing his own feed must accept a compensating loss of efficiency, calculate on capacity, and fortify the main ingredients of his ration to bring it up to accepted standards. Cereals and their by-products are the basis of all poultry feeds.

A mixture of as many as is practicable should be used. If that is supported by a variety of protein concentrates, which should include some of animal origin, there is a reasonable chance of meeting requirements of essential amino-acids. If the birds are not continually housed they are unlikely to be short of vitamins.

A supplement of calcium and phosphorus and 0.5 per cent of common salt is required for birds in high production.

The following are examples of rations which have proved to be satisfactory.

Marketing

The storage and transport problems associated with the marketing of poultry meat are the same as those associated with other kinds of meat but the marketing of fresh eggs presents some added difficulties. One of the chief causes of egg spoilage in the tropics is the development of the embryo in the fertile egg during the hottest months of the year.

In Malaysia

Ingredient	*Chick mash*	*Parts by weight* *Grower's mash*	*Layer's mash*
Yellow maize meal	40	53	30
Rice bran	24	24	20
Dried skim milk	20	—	—
Fish meal	10	10	24
Groundnut cake, decort.	—	10	10
Copra cake, double pressed	—	—	10
Dried brewers' yeast	3	—	—
Red palm oil	2	—	3
Bone flour or limestone	—	3	2
Common salt	1	—	1

In Ceylon

Ingredient	*Up to 2 months*	*Parts by weight* *3 to 4 months*	*Over 41 months*
Maize meal	35	37	38
Coconut meal	25	30	30
Fish meal	17	10	8
Rice bran (grade 1)	20	20	20
Skim-milk powder	2	1	—
Sterilized bone meal	—	½	1½
Oyster shell, ground	1	1	2
Common salt	—	½	½

Some people consider that the presence of an embryo in even an advanced stage of development adds to the value of the egg and, from the nutritional point of view, they are possibly correct but judged by other standards, a fertile egg which has been stored under hot tropical conditions is inedible after 48 hours.

The easiest way to avoid the difficulty is to exclude cocks from the laying flocks. When that has not been done, the development of the embryo can be stopped by immersing the eggs in water at 135° F. (57.2° C.) for 15 minutes as soon as they are collected, or they can be preserved by soaking them in lime water for about 18 hours or by treating them with white mineral oil of high viscosity. In any case palatability can be prolonged by reducing their temperature to at least 65°F. (19.4°C.) as soon as possible, and they can be preserved almost indefinitely when frozen.

Normal eggs do not decrease much in weight during storage in moderate atmospheric temperature but in a dry atmosphere if the air temperature is above 90°F. (32.2°C.) a loss of over ten per cent may occur and, in these circumstances, there may be difficulty in marketing eggs at a standard weight of two ounces (57 gm.).

DUCKS

Domestic ducks, like fowls, are raised throughout the tropics, but mostly in regions of high rainfall, in riverine areas and coastal districts. In India, for instance, 75 per cent of all the ducks in the sub-continent are in Bengal and the coastal regions of Madras.

Most breeds of domestic ducks are derived from the wild mallard, *Anus platyrhynchos*, but the Muscovy has its origin in the American *Cairina moschata*.

The total number of domestic ducks in the world is small compared to that of fowls and it may be assumed that it is the exacting nature of their climatic needs which restricts their suitability for domestic use, coupled with the fact that they eat more food, that neither their meat nor their eggs are so palatable, and that they foul their surroundings.

On the other hand, they lay more and bigger eggs than the indigenous fowl, they grow to a greater size, they require little housing, are susceptible to few contagious diseases, are better able to protect themselves from marauders and are excellent foragers.

In all suitable regions, and in many where the dryness of the climate would make it appear unsuitable, a few domestic ducks are found as part of the property of even the poorer householder but, particularly in tropical Asia, they are the sole source of livelihood of a considerable number of people who own huge flocks.

In tropical countries they are raised chiefly as egg producers and very seldom are they specially reared for meat. Ducks of some breeds such as the *Chinese* in Indo-China, the *Java,* in Malaysia, *the Indian Runner* and their crosses and the white-breasted *Nageswari* in India with an average body weight of no more than four pounds (1.8 kg.), lay on the average about 60 to 80 eggs per annum, each weighing about two ounces *(57 gm.),* while ducks of breeds such as the ubiquitous *Muscovy* and its crosses, the *Chinese Pekin* and many other indigenous breeds make good table birds weighing up to more than six-and-a-half pounds (2.95 kg.).

These results are attained from ducks which find all their own food or receive only the minimum supplements. Under intensive husbandry in both the tropics and in the temperate zone, Indian Runner ducks, for instance, lay as many as 300 eggs per annum, and at least one bird has laid 363 eggs in 365 days.

In wellmanaged flocks of the improved breeds the average annual production per bird is 200 eggs. The live body weight of the adult drake is four-and-a-half pounds (2.04 kg.). Under farm conditions an average of 150 eggs per bird per annum is expected from ducks of the East Asian egg-laying breeds. It is reported that a nondescript Philippine duck maintained in a battery laid 300 eggs in 365 days.

Drakes of the improved Pekin breed grow to an average weight of nine pounds (4.1 kg.) and the hens to eight pounds (3.6 kg.) which is the same average weight as the English *Aylesbury* and the French *Rouen*. Drakes of the Muscovy breed, under good conditions, average ten pounds (4.5 kg.) in weight but the ducks grow only to seven pounds (3.2 kg.).

The Indian Runner has been used extensively throughout the world in the evolution of high egg-producing breeds such as the *Campbell* with an average annual egg production of about 250. The Muscovy is frequently used to increase the size of ducks of other breeds or to produce large crossbreds from them.

Management

Any method of poultry-keeping outlined in the previous section is applicable to ducks, but the relative increase in output obtained by intensive husbandry is less than that from fowls and the methods may or may not be as profitable as well-established indigenous ones. Few methods of husbandry can be more economical than that of the specialized duck-breeders of South-east Asia.

Living in country where, during the breeding season, large areas of marsh and flood afford ample food for ducks, these people hatch thousands of eggs by the old Chinese method of artificial incubation. Ducks are more easily 'sexed' than fowls, and these specialists, after dividing their young stock at the earliest age into flocks of males and females, are ready to dispose of them at any time.

When the birds are old enough, traders move them on foot in huge flocks over long distances through well-known tracts, feeding them on the gleanings from paddy fields or such-like crops, and selling eggs and stock as they go.

In any circumstances, breeding stock is best maintained on free range, and where that is not possible, the semi-intensive system is the most suitable. Simple houses suffice for accommodation provided they are dry and cool.

A suitable house should be situated on a dry foundation and may consist of a well-thatched roof over a hard earth floor about six inches (15.2 cm.) above the surrounding ground and enclosed by frames of wire netting. A house with floor space of 12 × 8 feet (3.65 × 2.4 m.) will accommodate 60 ducks up to seven weeks old; some 22 square yards (18.4 sq. m.) of run should be provided.

For a large flock, the roof should stand eight feet (2.44 metres) high in front and fall in a span of 15 feet (4.57 metres) to a height of four-and-a-half feet (1.37 metres) behind. For a small breeding flock the roof need be no more than a few feet high in front.

Three square feet (0.279 square metres) of floor space is required for each adult bird. Wire-netting or slatted floors help to keep the accommodation clean and dry but, in any case, easily changed litter such as straw or dry reeds should be liberally strewn on the floor. Nest-boxes, each having a floor space of 12 by 16 inches (30.5 by 40.6 cm.) may be provided at ground level at the rate of one for five birds; their front containing-board should be no higher than six inches (15.2 cm.) but ducks often prefer the floor to a nest. Nests need not be provided for small flocks. Out-runs can with advantage include an area of swimming water but it is not essential.

Ducks usually lay early in the morning and those kept on open range are usually confined to their night quarters till the sun is well up and laying has been completed. If ducks are to be maintained on fresh grass-land they should be stocked at no more than 175 to the acre (0.40 hectare). In special circumstances laying ducks can profitably be kept in batteries.

Breeding

The usual number of ducks allowed to one drake is six but where fertile eggs are the main consideration, the number should be no more than four. The drakes should be put with the ducks at least one month before fertile eggs are needed. The belief that effective copulation can only take place in water is mistaken.

It is, however, desirable that ducks should have access to water to swim on as they are able to keep themselves clean, and nesting ducks are better able to keep their eggs at the correct humidity. Muscovy crossbreds such as the Muscovy × Java and the Muscovy × Manilla are sterile, but fertile crossbreds result from the mating of domestic ducks with such wild species as the Pintail and the Shoveller.

Ducks usually begin to lay at about six months of age if they are of high egg-producing strains. Ducks of the modern laying breeds are now for all practical purposes non-broody and even those of such breeds as the Pekin and Aylesbury cannot be relied on to incubate their eggs. The Java type, common throughout Southeast Asia, is also non-broody.

Hens will incubate duck eggs satisfactorily if the eggs are damped at least once daily. A hen can hatch twelve duck eggs. The incubation period is four weeks for all breeds except the Muscovy for which it is five weeks. Muscovys are capable of hatching up to 30 eggs but it is more usual for ducks to hatch 15 to 20.

When duck eggs are artificially incubated, a temperature lower by one degree Fahrenheit (0.56° C.) than that needed for hens' eggs should be maintained and it is most important that the relative humidity should never be allowed to drop below 60 per cent and it may with advantage be held at 70 per cent during the first 24 hours. From the second to the fourteenth day the eggs must be turned twice daily and sprinkled with tepid water at least once a day and after that three times daily.

Rearing

Ducklings hatched in an incubator should be transferred to a brooder as soon as they have dried. The 'cold' type of brooder is very satisfactory. If for any reason it is necessary to use artificial heat, a temperature of 85° F. (29.4°C.) is sufficient for the first night or two and it may be dropped gradually to 65° F. (18.3°C.) or to ambient temperature within a fortnight. Ducklings are generally remarkably active and should be allowed ample space in the open. If they have been hatched naturally they should be left with the duck for two or three weeks as they do better when brooded by her.

Feeding

Ducklings should be fed at least four times a day until they are two weeks old; afterwards, for another two weeks the number of feeds a day may be no more than three; after this, the number and size of the feeds can be gradually reduced as the bird's foraging power develops.

Three feeds a day must continue to be given to those in confinement until they are eight weeks old, but thereafter twice a day is sufficient except for those being fattened for killing.

Food recommended for chicks is suitable for ducklings. One-day-old birds should be given coarse milled cereals moistened with milk or water as a first feed and then fed on a mash composed on the following lines:

Cereals	30 parts
Fine cereal offal (bran)	30 "
Fish or meat meal	20 "
Extracted oil-cake meal	10 "
Fine grit	5 "

The mash should be damped just sufficiently to make it 'crumble'; if it is too wet much of it is lost through the sieving process to which it is subjected in the duck's bill. The food should always be given in flat-bottomed, shallow troughs and at each meal no more than what can be eaten in about ten minutes should be fed. Grit or sand should be available *ad lib.* Twenty-four running feet (7.3 m.) feeding space is required for 100 ducklings up to three weeks old, and 40 feet (12 m.) thereafter.

An adult duck in full lay can consume as much as ten ounces (283 gm.) of food each day, but the average amount is six ounces (170 gm.). Food mixtures recommended for fowls are suitable for ducks and are the ones generally used for birds kept under modern intensive conditions. A suitable mash for layers is composed of:

Cereals	75 per cent
Animal protein	10 "
Oil-cake meal	10 "
Dried edible fodder	5 "
Salt	0.5 "
Limestone	0.25 "

A typical mash is the one recommended by the College of Agriculture in the Philippines:

Rice bran	60 parts
Ground maize	20 "
Fish or shrimp meal	20 "

to which is added two per cent of cod-liver oil and two per cent ground oyster-shell for housed birds.

Under less intensive husbandry greater use can be made of local food-stuffs in which case up to one-third of the total diet may be composed of the cheapest vegetable foods or young edible fodder such as burseem, centrosema, or sweet-potato shoots.

Ducks on free range are kept in greatest numbers where fish and fish waste are available, and fish then forms the basis of their diet and is balanced by an evening feed of such cereals as rice, sorghum, maize and their by-products or by yams, sago and other carbohydrate feeds.

A plentiful supply of clean drinking water should always be available near their feeding troughs. The water should be deep enough to allow the birds to immerse their heads.

GEESE

Geese are raised primarily as meat producers. They are herbivorous and can be reared most economically where at all seasons a natural supply of fresh, young fodder is available for their food. This means that they are best suited for areas of high rainfall or wet marshy land.

Although certain wild species migrate to the tropics during the cool seasons of the year, no domestic breed has been developed in that zone and those evolved elsewhere do not easily adapt themselves to either arid conditions or very high temperatures.

Common breeds of domestic geese developed for modern requirements are: the Toulouse, Emden, African, Chinese, Canadian and Egyptian; in Britain, the English, Brecon Buff, Roman and Sebastopol are also well known.

The *Toulouse* and *Emden* are European : the former is grey and the latter white in colour. The birds of both breeds are large, grow quickly and have a propensity to fatten. The average weight of the full-grown gander is 26 and 20 pounds (11.8 and 9.1 kg.) respectively, and that of the goose 20 and 18 pounds (9.1 and 8.2 kg.).

Geese of the *African* breed closely resemble the above in weight, size and habits. They are grey in colour but are distinguished by a knob over the base of the bill and by a pronounced pouching of the throat. Under tropical conditions they do not grow to the same size as well-bred birds in the temperate zone.

The *Chinese* breed, of which there are two varieties, the Brown and the White, is found throughout the high-rainfall area of Southeast Asia. The average weight of a well-bred and fed gander is 12 pounds (5.4 kg.) and that of the goose 10 pounds (4.5 kg.) but the weight of the nondescript birds raised under ordinary village conditions is little more than half. This breed is also distinguished by a knob over the bill and a slight pouching of the throat.

On the other two breeds, the *Canadian* and the *Egyptian,* the latter is more suited to tropical conditions. It is some two pounds (0.91 kg.) lighter than geese of the modern Chinese breed but it is active, hardy and a good forager.

Management

Geese are not kept on a commercial scale in the tropics but are maintained only in small domestic flocks. They are invariably allowed free range without restriction where, under this system, they entail little trouble and no expense.

Doubtless, if their movements were restricted and they were well fed they could make rapid increas in weight as they do elsewhere, but it is improbable that gains as hig as ten pounds (4.5 kg.) in a month under forced feeding and six pounds (2.7 kg.) under less intensive treatment could be attained, as they are in the temperate zone. There is, however, no market for fat geese in the tropics, but large geese in good conditions are readily sold for slaughter.

Besides dry, well-ventilated shelter at night such as can, be given by a structure of simple, inexpensive design, protection muust be given from the sun during the day when it is at all intense.

Breeding

Geese may begin laying at one year old but should not be used for breeding until they are mature at three years old; the gander can be used when two years old.

Five geese are usually allowed to each gander but three geese of the heavy breeds are sufficient. Mating is usually for life asid difficulty may be encountered if geese are added to an established flock or if the gander is changed.

In good-laying strains of the Chinese breed up to 60 eggs may be expected per goose per annum, but the nondescript bird lays on the average about 20, the eggs being some four to five ounces (113 to 142 gm.) in weight. Birds of the heavy breeds lay from 30 to 40 eggs in a season. Egg-laying is highest during the third and fourth year of life but geese continue to lay satisfactorily up to about ten years of age.

Hens, being more amenable to management than geese, are often used for hatching the eggs of high-laying strains of breeds when the maximum egg production is of importance and where artificial incubators are not used.

A large hen is required and everl such a one can cover only four eggs of geese of the smaller breeds. The eggs must be turned by hand each day and liberally damped every day after the middle of the incubation period, which is 30 days for the light breeds and up to 34 for the others. A goose can cover up to 15 of her own eggs and no handling is required.

In an artificial incubator without forced air a temperature of between

101.5° and 102.5° F. (38.6 and 39.2° C.) just above the eggs is required; in forced-air incubators a temperature of 99° F. (37.2° C.) should be maintained. The humidity must be kept as high as that needed for ducks and the eggs must be turned, aired and sprayed twice daily up to the 28th day.

Rearing

Goslings hatched artificially should be treated in the manner recommended for ducklings but they are hardier, grow very fast and eat proportionately more food. Dry, clean accommodation during the early days is essential but only in exceptional circumstances is artificial heat required.

Feeding

Food should be offered to goslings when they are a day old, and should consist of coarsely milled cereals moistened with milk or water. After two days finely chopped, tender greens should be added to make up about one-third of the feed. For the first day or two they should be fed four or five times a day but at the end of a week, three times is sufficient.

If suitable pasture is available, that is, young, growing grass, it will provide all the food necessary for the birds from three to four weeks old. If the grazing is of inferior quality or quantity, a supplement of chopped greens and cereal offal, fed as a mash, is required. The amount of food which should be given at one meal is as much as will be eaten in about ten minutes.

Fresh drinking water should be constantly at hand from the time the bird is hatched. Water in which to swim, as with ducks, is desirable but not essential for good health and fertility.

When geese are to be fattened, they should be confined in a cool, dry, but rather dark house so that there is neither opportunity nor encouragement to exercise. They should be fed on inexpensive carbohydrate foods such as sweet-potato, yam, rice or maize, to appetite, for about a month.

TURKEYS

The domestic turkey is derived from the American wild turkey of which there are two kinds; the Ocellated turkey, a native of the Yucatan Peninsula and thus a truly tropical bird, and the North American. It is from the latter that the modern domestic turkey has been developed. There are now many varieties, some standardized and others more or less in the process of becoming so.

The *Mammoth Bronze* typifies the large variety and is by far the most popular breed because not only does it attain a bigger size as an adult but it reaches a greater weight at an earlier age than any other breed; moreover, it is an excellent forager and very hardy.

The *Austrian White* is comparatively small but its meat is considered to be of better quality than that of the Mammoth, it is more domesticated in its habits and tolerates confinement better. From these two breeds the modern *Broad Breasted Bronze* and the *Beltsville Small-type White* have been evolved.

The former is even heavier than the Mammoth, carrying as it does a fleshy breast and thighs; the latter by the superior quality of its flesh and its restricted size complies more with the requirements of a domestic bird for family use both during its life and after it has been killed. The average live-weight of the adult Broad Breasted Bronze hen is 20 pounds (9.1 kg.) and that of the Beltsville White is 13 pounds (5.9 kg.); the weight of the cocks is 41 and 25 pounds (18.6 and 11.3 kg.) respectively.

Comparatively few turkeys are found in tropical countries in the Old World. It is said that they are too expensive to maintain, lay too few eggs and are slow to mature. The lighter breeds do very well in dry, hot climates when kept at liberty and provided with shade.

When given roomy, dry accommodation, they can be reared without much difficulty in hot climates where the rainfall is not excessive. They are foragers by nature and they thrive best where they can rove about while feeding on seeds, insects and young green grass and herbage. Under such conditions a stag with twelve breeding hens should be allowed an area of about one acre (0.4 hectare).

When they have to be confined, the accommodation which appears to be best suited to their needs is an open yard in which cover is provided by a well-thatched roof supported eight feet (2.44 metres) above the ground by metal uprights and, where protection against rain or wind is needed, with walls on the exposed sides.

A floor space of ten square feet (0.93 square metres) is required for each adult bird of the light breeds and 14 square feet (1.3 square metres) for those of the heavy breeds. Perches, which should be three inches by one inch (7.6 by 2.5 cm.) for poults and four by two inches (10.2 by 5.1 cm.) for adults, should be two feet (0.61 metre) from the ground and some two feet (0.61 metre) apart. The whole structure must be wired-in to provide protection from marauders.

Movable folds similar to those described for fowls but appropriately slatted, are also suitable. The minimum size should be 18 by 7 feet (5.5

by 2.1 metres) providing accommodation for one stag with his hens or twice that number of poults.

Breeding

Turkeys may start to breed when they are one year old but they are not fully mature until they are between two and three years old. Birds of the improved breeds, in temperate climates lay, on the average, up to 90 eggs per annum but, even in well-established flocks, there is great variation in the number of eggs laid by each hen.

The average weight of the egg is just over three ounces (85 gm.). The 'pause' in the rhythm of production, that is, the interval which periodically breaks the sequence of egg-laying days, may be pronounced as it is sometimes of several days' duration. The nondescript type of turkey of which the domestic flock in the tropics is so largely composed, lays a small egg weighing just over two ounces (57 gm.) and she seldom lays as many as 20 before going broody.

Up to ten hens are allowed each stag or male turkey. Stags tend to be selective in their mating to the detriment of the favoured hen and of the flock, although only one effective copulation suffices to fertilize a whole clutch of eggs and thus is required only every three weeks or so. The size of the male may be so disproportionate to that of the female that the hen may be badly damaged by him especially if his spurs are long and sharp.

Turkeys prefer to make their own nest but they can be induced to lay where it is most convenient by the provision of roomy, wellprotected nests. A suitable nesting shelter design for birds on range is the wigwam type set below three poles covered on the two sides by thatch, matting or sacks.

The incubation period is 25 to 28 days. The hens of the small tropical variety are good brooders; those of the larger type are clumsy and somewhat unreliable. In any case, if the eggs are not to be incubated artificially, it is generally more satisfactory to have them hatched under local fowls, and the turkey will then proceed to lay her next batch of eggs.

The precautions regarding the handling and storage of eggs and the management during artificial incubation advised for fowls, apply to turkeys, but more attention has to be paid to the maintenance of atmospheric moisture. The relative humidity should be 63 per cent with the dry-bulb thermometer at 99° F. (37.2° C.) and, after the 24th day, it should be 70 to 79 per cent at a dry-bulb temperature of 97° F. (36.1° C.). In incubators with forced air current, the temperature should be kept at 99.5° F. (37.5° C.) until the 24th day when it should be lowered to 97° F. (36.1° C.). In machines with unassisted ventilation, a temperature of 102° F. (38.9° C.) should be maintained one-and-seven-eighth inches (4.76 cm.) above the eggs. The eggs should not be turned during the first day of incubation or after the 24th day.

Rearing

Brooding methods suitable for chicks are suitable for poults but the latter, when very young, are more readily affected by climatic temperature and must be more carefully protected from direct rays of the sun as well as from a sudden fall in temperature such as occurs at night. They are peculiarly susceptible to dampness, especially if it is associated with cold.

During brooding the optimum temperature is 95° F. (35° C.) to 100° F. (37.8° C.) until the poults are a week old, after which it should be lowered by 1° F (0.6° C.) each day until the normal ambient temperature is reached.

Young turkeys are very susceptible to parasitic infestation as well as to most of the bacterial and virus diseases of poultry. Poults should be kept from contact with grown birds and chickens. 'Blackhead', a devastating disease of young turkeys, is carried by a common parasitic nematode, and so if they are confined they should never be crowded, should have good ventilation and should be on a wire or slatted floor.

Twelve square inches (77.4 sq. cm.) of ground space are required in the 'hoover' for each poult. In the brooder a square foot (0.093 square metre) is needed for each bird up to six weeks old; four square feet (0.372 square metre) till twelve weeks, and eight (0.744 square metre) till they are 24 weeks old.

A small brooderhouse eight feet (2.44 metres) square, four feet (1.22 metres) high in front and three-and-a-half feet (1.07 metres) behind, placed in a wired-in fold 20 yards (18.3 metres) square will amply accommodate a batch of 20 poults until they are about ten weeks old.

Feeding

A peculiarity about turkeys is their aversion to any change in their feeding routine or in the nature of their food. Poults have very poor sight and those raised artificially may sometimes appear to have no desire to eat and either die or be slow in acquiring an appetite and in gaining strength.

At one day old they should be offered and encouraged to drink and to eat small quantities of fine mash, damped with milk when it is available, or should be given finely broken grain and, until they begin to forage, it is usual to mix in finely chopped, tender, green food.

Any of the foods advised for fowls is suitable but the protein content should be somewhat higher, that is, about 23 per cent, up to the time when the head appendages of the young birds develop at about ten weeks, and should be gradually reduced to about 15 per cent when they are adults.

Thus, if one of the mixtures designed as a starting diet for chicks is used, it should be supplemented by about three to four per cent of a food rich in animal protein, such as fish meal. For adults, in addition to the usual kind of laying mash recommended for fowls, two to four ounces (56.7-113.4 gm.) of a cereal grain should be fed daily.

The food capacity of adults of the light breeds is only four ounces (113.4 gm.) a day, but that of the heavy breeds such as the Broad Breasted Bronze is about ten ounces (283.5 gms.).

A supply of green food, flint grit and lime should always be available if the mash has been compounded without due allowance for such supplements. Turkeys are as sensitive as domestic hens to nutritional deficiencies.

It takes fifty pounds (22.68 kg.) of well-balanced mash to produce a fourteen-pound (6.35 kg.) turkey at twenty-four weeks, provided the turkey has been healthy, is of a good strain and has been reared under good conditions.

INDEX

N

O

P

T

U

V

W

Z